高等学校专业英语教材

科技前沿英语读写教程

主编 王 宇 李映夏 朱必可
副主编 高 莹 杜宛宜 周纯岳 刘 辉

电子工业出版社
Publishing House of Electronics Industry
北京·BEIJING

内 容 简 介

本书分为8章，每章围绕一个特定的科技主题展开讨论，内容涵盖了当前多种前沿技术。Chapter 1 介绍如何利用 6G 技术实现全息影像社会和可触摸网络。Chapter 2 介绍纳米芯片技术的突破和存在的挑战。Chapter 3 关注以生成对抗网络为基础的图像生成技术。Chapter 4 介绍智能机器人技术的发展。Chapter 5 介绍智能电动汽车技术取得的成果。Chapter 6 聚焦火箭回收技术的突破和存在的挑战。Chapter 7 讨论云计算技术在能耗方面的问题。Chapter 8 介绍智慧农业技术带来的便利和挑战。

除了介绍技术，本书每章都设计了一个中国专题，介绍中国在这些技术领域取得的成就和突破，并围绕该专题进行讨论，以帮助学生更好地将专业知识与前沿科技相结合，坚定科技报国的信念和决心。

本书适合信息技术相关专业本科二年级以上的学生使用，有一定的难度，但能够帮助学生更深入地了解和学习前沿科技领域的最新进展。

未经许可，不得以任何方式复制或抄袭本书之部分或全部内容。
版权所有，侵权必究。

图书在版编目（CIP）数据

科技前沿英语读写教程 / 王宇，李映夏，朱必可主编． -- 北京：电子工业出版社，2024.12． -- ISBN 978-7-121-49652-3

Ⅰ．G301

中国国家版本馆 CIP 数据核字第 2025ES3447 号

责任编辑：刘 瑀
印　　刷：三河市龙林印务有限公司
装　　订：三河市龙林印务有限公司
出版发行：电子工业出版社
　　　　　北京市海淀区万寿路 173 信箱　邮编：100036
开　　本：787×1092　1/16　印张：12　字数：399 千字
版　　次：2024 年 12 月第 1 版
印　　次：2024 年 12 月第 1 次印刷
定　　价：49.00 元

凡所购买电子工业出版社图书有缺损问题，请向购买书店调换。若书店售缺，请与本社发行部联系，联系及邮购电话：(010) 88254888，88258888。
质量投诉请发邮件至 zlts@phei.com.cn，盗版侵权举报请发邮件至 dbqq@phei.com.cn。
本书咨询联系方式：liuy01@phei.com.cn。

前　　言

在人工智能（AI）技术大爆发的背景下，科技不断发展，前沿技术层出不穷。本书就是结合前沿科技编写的，旨在帮助学生深入了解和学习前沿科技领域的最新进展。

本书注重培养学生的思辨能力，通过阅读和写作让学生深入思考，并用英语表达自己的观点。此外，本书展示了由 ChatGPT 和 ERNIE Bot（文心一言）生成的观点，引导学生思考人工智能产出观点的利弊，并正确使用大语言模型。本书采用以口语表达和文章写作产出为导向的教学方式，引导学生阅读前沿的科技信息，并掌握阅读技巧和专业术语。

本书具有以下特色：

- **前沿科技内容覆盖广泛**：本书内容涵盖当下热门的科技领域，包括人工智能、云计算、智能机器人等，使学生能够及时了解和掌握科技发展动态。
- **专业术语深入解读**：本书对前沿科技内容相关的专业术语进行了深入解读和讲解，帮助学生理解并熟练运用这些术语，提升专业素养和沟通能力。
- **设立中国专题**：本书每章都设立了中国专题，介绍中国在相关科技领域的成就和突破，增强对中国科技发展的了解。
- **思辨能力培养**：本书注重培养学生的思辨能力，通过阅读和写作等练习，引导学生深入思考和分析问题，并用英语表达自己的观点，提升其批判性思维和逻辑思维能力。
- **实践性教学**：本书以口语表达和文章写作产出为导向，让学生在实践中学习，通过大量的阅读、讨论和写作活动，提升英语水平和表达能力。
- **引入人工智能技术**：本书展示了 ChatGPT 生成的观点，让学生了解人工智能在教学中的应用，同时引导学生正确使用大语言模型。
- **多媒体辅助学习**：教材配有与主题相关的视频，为学生提供多维度、更具象的学习体验，帮助他们更好地理解课文内容和科技主题。

本书由大连理工大学王宇、李映夏、朱必可担任主编，高莹、杜宛宜、周纯岳、刘辉担任副主编。尽管我们做了最大努力，本书难免有不足之处，请广大读者批评指正。

编　者

目　录

Chapter 1　6G ·· 1
　Before You Read ·· 1
　Text ·· 2
　Reading Comprehension ·· 5
　Language Building ·· 8
　Critical Reading and Writing ·· 13
　Discussion and Presentation ··· 16
　Video ·· 19
　Summary and Reflection ·· 20

Chapter 2　Chip ·· 22
　Before You Read ·· 23
　Text ·· 25
　Reading Comprehension ·· 27
　Language Building ·· 31
　Critical Reading and Writing ·· 36
　Discussion and Presentation ··· 39
　Video ·· 42
　Summary and Reflection ·· 43

Chapter 3　AI Image Generation Technology ·· 46
　Before You Read ·· 46
　Text ·· 48
　Reading Comprehension ·· 51
　Language Building ·· 54

Critical Reading and Writing ·· 60
　　Discussion and Presentation ·· 62
　　Warm Up ··· 64
　　Video ·· 65
　　Summary and Reflection ·· 66
Chapter 4　Robot ·· 68
　　Before You Read ··· 68
　　Text ·· 70
　　Reading Comprehension ·· 72
　　Language Building ··· 75
　　Critical Reading and Writing ·· 79
　　Discussion and Presentation ·· 83
　　Video ·· 87
　　Summary and Reflection ·· 88
Chapter 5　Electric Vehicle ·· 91
　　Before You Read ··· 91
　　Text ·· 92
　　Reading Comprehension ·· 95
　　Language Building ··· 99
　　Critical Reading and Writing ·· 106
　　Discussion and Presentation ·· 109
　　Video ·· 113
　　Summary and Reflection ·· 115
Chapter 6　Aerospace ·· 117
　　Before You Read ··· 117
　　Text ·· 118
　　Reading Comprehension ·· 120
　　Language Building ··· 123
　　Critical Reading and Writing ·· 129
　　Discussion and Presentation ·· 132
　　Video ·· 134
　　Summary and Reflection ·· 136

Chapter 7　Cloud Computing ·· 138

　　Before You Read ··· 138

　　Text ··· 140

　　Reading Comprehension ·· 141

　　Language Building ·· 145

　　Critical Reading and Writing ·· 151

　　Discussion and Presentation ··· 155

　　Video ··· 158

　　Summary and Reflection ·· 159

Chapter 8　Smart Farm ·· 162

　　Before You Read ··· 162

　　Text ··· 164

　　Reading Comprehension ·· 166

　　Language Building ·· 169

　　Critical Reading and Writing ·· 174

　　Discussion and Presentation ··· 177

　　Video ··· 180

　　Summary and Reflection ·· 181

Chapter 1
6G

> **Objectives**
> In this chapter, you should be able to:
> - Analyze and discuss cutting-edge technological advancements of the emergence and potential applications of 6G technology.
> - Develop critical thinking, reading, and writing skills by evaluating the feasibility and implications of 6G technology in various scenarios.
> - Demonstrate the ability to summarize and categorize information from complex texts related to 6G technology using techniques such as multiple choice, mind mapping, and matching exercises.
> - Expand vocabulary and language proficiency in an academic context through exercises like glossaries, word choice, and translation tasks related to 6G technology and its terminology.
> - Engage in collaborative discussions and presentations to share insights and perspectives on 6G technology and its implications, both before and after reading relevant material.

 ## Before You Read

A. Discussion
Look at the pictures below and discuss with a partner.
1. What technologies are applied in the three different scenarios?
2. How are they related to 6G?

B. Skimming and Scanning

Browse the text and answer the questions below.

1. Which of the followings is NOT the characteristic of 6G technology?

A. High fidelity

B. High frequency

C. Low energy

D. Low latency

2. What is the speed of chip-to-chip communication?

A. 100 Gbps or more

B. 150 Gbps or more

C. 200 Gbps or more

D. 250 Gbps or more

3. What can NOT be realized in the "high-fidelity holographic society"?

A. Remote surgeries

B. Remote education

C. Remote repair

D. Remote dining

 Text

Study: 6G's Haptic, Holographic Future?

Possibilities and Challenges for Future 6G Communications Networks

1. Imagine a teleconference but with holograms instead of a checkerboard of faces. Or envision websites and media outlets across the Internet that allow you to make haptic connections. Researchers studying the future of sixth-generation (6G) wireless

communications are now sketching out possibilities—though not certainties—for the kinds of technologies a 6G future could entail.

2. Sixth-generation wireless technology—says Harsh Tataria, a communications engineering lecturer at Lund University, Sweden—will be characterized by low latency and ultrahigh frequency, with data transfer speed potentially hitting 100 Gbps. Tataria, along with colleagues from Lund University, Spark New Zealand, University of Southern California (USC), and King's College London, recently published a paper in Proceedings of the IEEE, presenting a holistic, top-down view of 6G wireless system design. Their study began by considering the challenges and technical requirements of next-generation networks, and forecasting some of the technological possibilities that could be practically realizable within that context.

3. Such future-casting is to be expected as 5G deployment picks up speed around the world, at which point subsequent generations of wireless technologies come more into focus. Tataria calls this "a natural progression", to look at the emerging trends in both technology and consumer demands. "When we look at 6G, we're really looking at vastly connected societies," he says, "even a step beyond what 5G is capable of doing, such as real-time holographic communications."

4. The study outlines what it calls a "high-fidelity holographic society", one in which "Holographic presence will enable remote users to be represented as a rendered local presence. For instance, technicians performing remote troubleshooting and repairs, doctors performing remote surgeries, and improved remote education in classrooms could benefit from hologram renderings." The authors note that 4G and expected 5G data rates may not enable such technologies—but that 6G might—owing to the fact that "holographic images will need transmission from multiple viewpoints to account for variation in tilts, angles, and observer positions relative to the hologram."

5. Even simple phone conversations could involve new levels of multimedia-rich experience. "For example, in this interview…we could be talking to the rendered presence of each other," says Mansoor Shafi, another study co-author. "and that would provide a much richer experience than the audio call we are having at the moment."

6. Another promising possibility the study teases involves what they call a haptic Internet.

"We believe that a variety of sensory experiences may get integrated with holograms," the authors write. "To this end, using holograms as the medium of communication, emotion-sensing wearable devices capable of monitoring our mental health, facilitating social interactions, and improving our experience as users will become the building blocks of networks of the future."

7. Mischa Dohler, another co-author, believes that 6G will consolidate the "Internet of skills" or the ability to transmit skills over the internet. "We can do it with audio and video, but we can't touch through the Internet…or move objects." The consolidation of edge computing, robotics, AI, augmented reality and 6G communications will make this possible, he says. "This next generation Internet…will democratize skills the very same way as the Internet has democratized information."

8. Molisch also hopes that 6G will bring better chip-to-chip communication. "As we go to 200 Gbps or more…the cable connections are just not able to keep up," he says. "As we are 'moving' to higher data rates, higher processing speeds…wireless links are one way in which this bottleneck can be overcome." This also means increased reliability as wireless connections are not impacted by shaking or vibration, and lower costs because replacing cables "might be more expensive than just putting in wireless transceivers onto the chips."

9. Other use cases mentioned in the paper involve what they call extremely high-rate "information showers"—hotspots where one can experience terabits-per-second data transfer rates—mobile edge computing, and space-terrestrial integrated networks. But, as Molisch cautions, "There is still a lot of research that needs to be done…before the actual standardisation process can start."

10. With 6G going up to terahertz frequencies, there will be tremendous challenges in building new hardware as well, the researchers say. Better semiconductor technologies will also be needed for faster devices. Other challenges remain as well, including power consumption.

11. With frequency bands moving up in the hundreds of gigahertz, "even fundamental things like circuits and substrates to develop circuits are extremely tricky," says Tataria. "So getting all those things right, and going from the fundamental-level details all the way up to building a system is going to be substantially harder than what it first came across."

Their study, therefore, attempts to explore the trade-offs involved in each futuristic technology.

12. As the authors point out, this study is not a comprehensive or definitive account of 6G's capabilities and limitations—but rather a documentation of the research conducted to date and the interesting directions for 6G technologies that future researchers could pursue.

 Reading Comprehension

A. Multiple Choice
Choose the best answer for each question.

1. What characterizes sixth-generation (6G) wireless technology according to Harsh Tataria?
   A. Low latencies and ultrahigh frequencies
   B. High latencies and low data transfer speeds
   C. Low frequencies and high latencies
   D. High latencies and ultrahigh data transfer speeds

2. According to the article, what is a potential benefit of a "high-fidelity holographic society" enabled by 6G technology?
   A. Improved remote education in classrooms
   B. Enhanced audio quality in phone conversations
   C. Reduced need for edge computing
   D. Increased reliance on cable connections

3. What is one of the promising possibilities mentioned in the study regarding the integration of sensory experiences with holograms?
   A. Remote surgeries performed by holographic doctors
   B. Emotion-sensing wearable devices for monitoring mental health
   C. Improvement of traditional phone conversations
   D. Integration of holograms with augmented reality devices

4. According to Mischa Dohler, what is one aspect that 6G technology could

democratize?

A. Access to information

B. Access to skills

C. Access to medical services

D. Access to financial markets

5. What challenge is highlighted in the article regarding the development of sixth-generation (6G) wireless technology?

A. Decreased power consumption

B. Increased reliance on cable connections

C. Difficulty in building new hardware for terahertz frequencies

D. Decreased need for faster semiconductor technologies

B. Mind Map

How many main parts do you think the article is composed of? Group the paragraphs and fill in the blanks with the information you read from the article.

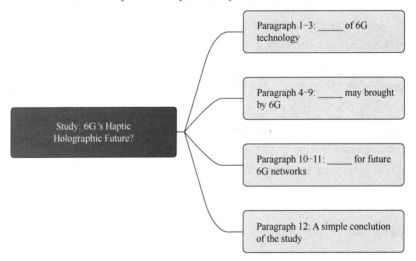

C. Matching

Read the text and decide which paragraph mentions the following information? Write the number of the paragraph before each sentence.

_____ 1. Because the frequency of 6G has risen to terahertz, it will encounter great difficulties in developing new hardware, from the most basic circuits and substrates to building systems.

_____ 2. Power consumption is also a huge challenge.

_____ 3. Tataria, together with other colleagues published a paper which presented a holistic view of 6G wireless system design.

_____ 4. 6G, short for sixth-generation wireless technology is being studied and analyzed by researchers.

_____ 5. The study is a documentation of the research to date and provides the direction for future research.

_____ 6. The natural progression towards 6G as 5G deployment accelerates globally

_____ 7. Holographic presence enables remote users to be represented locally, facilitating applications like remote troubleshooting, surgeries, and education.

_____ 8. Wearable devices will become the foundations to realize a haptic internet of the future.

D. Cloze

The information below is a summary of the text. Complete the summary by filling in the blanks with the words provided.

A. science	F. system
B. emotion-sensing	G. high-fidelity
C. semiconductor	H. terahertz
D. reality	I. haptic
E. communications	J. Internet

It's fascinating to delve into the potential of sixth-generation (6G) wireless communications as outlined in the study you provided. The vision of a "(1)_____ holographic society" and the concept of a "(2)_____ Internet" are particularly intriguing. It seems like 6G could usher in a new era of connectivity, enabling experiences and applications that were previously only imaginable in (3)_____ fiction.

The idea of remote surgeries conducted via holographic representations and the concept of transmitting skills over the internet, creating an "(4) _____ of skills," highlight the transformative potential of 6G. Additionally, the integration of (5)_____ wearable devices with holograms could revolutionize how we interact with technology and each other.

However, it's crucial to acknowledge the challenges that come with realizing this vision. Building hardware capable of operating at (6)_____ frequencies presents significant engineering hurdles, as does ensuring low power consumption and developing faster (7)_____ technologies. Moreover, the complexity of designing systems that operate at such high frequencies requires careful consideration of trade-offs and challenges at both the fundamental and (8)_____ levels.

While the possibilities outlined in the study are exciting, it's important to remember that much research and development are still needed before these technologies can become a (9)_____. Nonetheless, the study provides a valuable starting point for future exploration and innovation in the field of 6G wireless (10)_____.

 Language Building

A. Glossary

Proper Nouns
6G 第六代移动通信技术，一个概念性的无线网络移动通信技术，也称为第六代移动通信技术，目的是促进互联网的发展。6G 仍在开发阶段，其传输能力可能是 5G 的 100 倍，网络延迟也可能从毫秒级降到微秒级。
Lund University 隆德大学，是瑞典一所现代化、具有高度活力和历史悠久的学校，也是综合研究型大学和世界百强大学。
Spark New Zealand 一家新西兰电信公司。
University of Southern California（USC） 南加利福尼亚大学，又称为南加州大学，简称南加大，是美国西海岸私立研究型大学，位于美国洛杉矶市。
King's College London 伦敦国王学院（简称 King's 或 KCL），是位于英国伦敦的一所公立综合研究型大学。

Proceedings of the IEEE

 Proceedings of the IEEE 是刊载电子、电气工程和计算机科学技术发展相关述评、调查和论文的领先期刊。

Academic Words	
hologram (n.)	全息图（激光）
envision (v.)	想象，预想
haptic (a.)	触觉的
sketch (v.)	简述，概述
latency (n.)	延迟
frequency (n.)	频率，频次
holistic (a.)	整体的，全面的
progression (n.)	发展，前进
emerging (a.)	新兴的，发展初期的
fidelity (n.)	精确性，保真度
remote (a.)	远程的
render (v.)	渲染
transmission (n.)	传送，发送
democratize (v.)	普及，使大众化
cable (n.)	电缆；钢缆，缆绳
bottleneck (n.)	瓶颈；障碍物
reliability (n.)	可靠性；可信度
terabit (n.)	兆兆位（量度信息的单位）
terrestrial (a.)	陆地上的，地面上的
integrated (a.)	综合的；集成的
standardization (n.)	标准化
terahertz (n.)	太赫兹，兆兆赫兹（频率单位）
tremendous (a.)	巨大的，极大的；极好的
consumption (n.)	消费，消耗
gigahertz (n.)	千兆赫
circuit (n.)	电路，回路
substrate (n.)	基板，基底
substantially (adv.)	大幅度地
futuristic (a.)	未来的

B. Words and Phrases

Exercises 1   Word Choice

Use the words in the box to finish the sentences.

haptic	holistic	subsequent	rendering	promising
monitor	democratize	vibration	transceiver	edge

1. No computer is as smart as a human being with a _____ point of view.
2. Players use a _____ device such as a joystick to control the game.
3. It automatically switches off the sound and puts itself into _____ mode.
4. _____ studies have come up with similar results.
5. AI looks at resumes in greater numbers than humans are be able to and selects more _____ candidates.

Exercise 2   Phrases

Match the words provided below with appropriate one in the box.

1. _____ network
2. edge _____
3. _____ communication
4. _____ transceiver
5. _____ Internet

| holographic |
| wireless |
| haptic |
| integrated |
| computing |

Exercise 3   Sentence Completion

Complete the sentences by filling in the blanks with phrases in the above exercise.

1. Cloud computing and _____ are often used for comparison.
2. The enhanced _____ with built-in antenna makes it a step towards node-to-node communication.
3. One of the biggest challenges of the _____ is creating a sense of pressure on the skin without a physical surface.
4. The construction of the _____ system is a basic project for developing an automatic, netting, digitized library.
5. With the rapid development of 5G _____, holographic video calling, which appeared in science fiction movies in the past, will also enter the 5G era.

Exercise 4   Translation
Translate the sentences by using the words and phrases you have learned in the above exercises.

1. 它有可能极大地加快基因测序的速度和普及程度。
_____

2. 市场上已经出现很多设备，据说这些设备可以监测和帮助睡眠。
_____

3. 他说没人会找到我们，除非那个收发器好用了。
_____

4. 计算机信息处理技术的未来似乎将发生巨大的变化。
_____

5. 他当选为最有前途的新导演。
_____

C. Collocation

Exercise 1   Matching
Match the English phrase with the correct Chinese definition.

sketch out possibility	A. 技术上的可能性
technological possibility	B. 排除可能性
promising possibility	C. 低估可能性
theoretical possibility	D. 极有可能
rule out possibility	E. 有前景的可能性
distinct possibility	F. 简述出可能性
discount possibility	G. 理论上的可能性

Exercise 2   Blank Filling
Scan the text and complete the sentence containing the word "possibility".

1. Their study began by considering the challenges and technical requirements of next-generation networks—and forecasting some of the _____ possibilities that could be practically realizable within that context.

2. What had seemed impossible now seemed a _____ possibility.

3. The risks remain and we cannot _____ the possibility of further strikes.

4. Researchers studying the future of sixth-generation (6G) wireless communications are now _____ possibilities—though not certainties—for the kinds of technologies a 6G future could entail.

5. He has not _____ the shock possibility of a return to football.
6. This is a _____ possibility, but it is difficult to implement.
7. Another _____ possibility the study teases involves what they call a haptic Internet.

Exercise 3   Translation
Translate the sentences below from Chinese to English using "possibility" and its collocation in this section.
1. 医生会进行身体检查、面谈和实验室检测，以排除这些可能性。
_____
2. 这是一种理论上的可能性，但我认为这种情况不会发生。
_____
3. 过去看似不可能的事，现在显然有可能了。
_____
4. 虽然这不可想象，但我认为我们不应低估这种可能性。
_____
5. 这份提纲简述了实行这个项目的可能性。
_____

D. Terminology
Exercise 1   Table Filling
Read the article and find the English technical terms according to the Chinese equivalents.

English Technical Terms	Chinese Equivalents
	实时全息通信
	渲染技术
	无线收发器
	移动边缘计算
	集成网络
	触觉互联网

Exercises 2 Blank Filling
Use the terms in the above exercise to complete the sentences below.
1. "When we look at 6G, we're really looking at vastly connected societies," he says, "even a step beyond what 5G is capable of doing, such as _____."

2. "For example, in this interview…we could be talking to the _____ presence of each other," says Mansoor Shafi, another study co-author.

3. Another promising possibility the study teases involves what they call a _____.

4. This also means increased reliability as wireless connections are not impacted by shaking or vibration, and lower costs because replacing cables "might be more expensive than just putting in _____ onto the chips."

5. Other use cases mentioned in the paper involve what they call extremely high-rate "information showers"—hotspots where one can experience terabits-per-second data transfer rates—_____, and space-terrestrial _____

Critical Reading and Writing

A. Brainstorming

Work in groups. Fill in the table according to the instruction.

List at least three industries that will be impacted by 6G technology, summarize your reasons and put each in one sentence.

Industries	Reasons
1	1
2	2
3	3

B. Critical Reading

Exercise 1: The following are the answers to the question provided by ChatGPT. Read the introduction, and then match the headings and the corresponding content in the body part.

Heading A: Traditional Broadcast Media

Heading B: Education and Training Services

Heading C: Traditional Retail

Heading D: Wired Communications Providers

Heading E: Physical Security Services

ChatGPT

The advent of 6G technology is likely to bring profound changes across many industries, but some are particularly susceptible to transformation or even obsolescence. Here are a few industries that might be significantly impacted:

a. _____:

With 6G's promise of ultra-high speed, low latency, and enhanced reliability, the need for physical wired connections (like broadband and telephone lines) could diminish. Wireless technology might become so reliable and fast that traditional cable and DSL services become obsolete.

b. _____:

As 6G technology enables even faster and more reliable streaming services, traditional broadcast mediums like radio and television could face further decline. People might choose streaming platforms that offer more versatile and personalized content delivered directly to mobile devices or home entertainment systems without the need for traditional broadcasting infrastructure.

c. _____:

With the integration of advanced AI, augmented reality, and ultra-reliable, low-latency, communications, physical security services (like patrolling guards) could be partially replaced by automated surveillance systems, drones, and remote monitoring services.

d. _____:

Enhanced virtual and augmented reality experiences powered by 6G could further push consumers towards online shopping, reducing the demand for physical retail spaces. Virtual try-ons, immersive virtual stores, and other advanced features could make online shopping even more convenient and attractive.

e. _____:

The potential for high-fidelity holographic presentations and interactions could revolutionize education and training. Traditional in-person training sessions and classroom learning could be replaced by remote, holographic instructors and trainers, leading to a decline in physical educational institutions and training facilities.

While it's important to note that these transformations do not necessarily mean these industries will vanish overnight, they could evolve significantly as they adapt to new technologies introduced by 6G. Each industry might find ways to incorporate 6G technology into their operations or pivot their business models to remain relevant.

Exercise 2: Do you agree with ChatGPT on the above ideas? Which one do you agree most? Which one the least? Share your ideas with the class.

C. Essay Writing

Having engaged in discussions and vocabulary preparation, you've likely generated numerous insightful ideas. Now, it's time to reflect on these ideas and the knowledge you've acquired by crafting an essay. Use the following instructions to guide your writing process:

Topic

The Impact of 6G Technology on … (a specific industry)

Background Information

In recent years, technological advancements have revolutionized various aspects of our lives, from communication and entertainment to healthcare and transportation. One of the most anticipated technological developments on the horizon is the advent of 6G technology. Building upon the capabilities of its predecessors, 6G is expected to offer unprecedented speed, reliability, and connectivity, ushering in a new era of innovation and transformation across industries.

As the world eagerly awaits the rollout of 6G technology, speculation abounds regarding its potential impact on various industries. From manufacturing and healthcare to education and entertainment, virtually every sector stands to be influenced in some way by the introduction of this cutting-edge technology. However, one question looms large: which industry will be most profoundly affected by the advent of 6G?

Instructions

In an essay of approximately 300-450 words, <u>analyze which industry could be mostly influenced by the advent of 6G technology</u>. Provide specific evidence and arguments to support your opinion.

Your essay should include the following components:

Introduction (approximately 50-75 words): Briefly introduce the topic and provide context for your analysis.

- <u>State your thesis or main argument regarding which industry you believe will be</u>

most impacted by the introduction of 6G technology.

Body Paragraphs (approximately 250-350 words): Present your argument in detail, supported by specific evidence and examples in 3 separate paragraphs.
- Consider factors such as the unique capabilities of 6G technology, current trends in the industry;
- Consider potential implications for stakeholders;
- Discuss how 6G technology could revolutionize operations, business models;
- Discuss consumer experiences within the chosen industry.

Conclusion (approximately 50-75 words): Summarize your main points and restate your thesis in light of the evidence presented.
- Reflect on the broader implications of your analysis
- Consider potential avenues for future research or exploration.

* Ensure that your essay is well-structured, logically organized, and supported by evidence from reputable sources. Use clear and concise language, and proofread your work carefully for grammar, punctuation, and spelling errors.

Discussion and Presentation

A. Group Discussion

Exercise 1: Think and discuss the questions below.

1. How might the disparity in R&D spending between the United States and China impact global technological leadership in the coming years, particularly in areas such as semiconductor chips, AI, and wireless technology?

2. As China advances rapidly in the development and testing of 6G technology, what implications might this have for the global technological landscape, including economic competitiveness, innovation, and geopolitical power dynamics?

3. Considering the potential societal impacts of 6G technology, such as increased automation and the transformation of work in various industries, how can societies prepare to address the challenges and opportunities presented by such advancements, particularly in terms of employment, healthcare, and governance?

Exercise 2: Read the following article. Search for any information related to the above questions. Do you think the answers provided in the article match yours? Share your ideas with the class.

B. Extended Reading-6G in China

Read the article, highlight any ideas or language that will help you in your presentation in the next session.

6G: A Case for Why China Will Become More Innovative

The U.S. National Science Foundation (NSF) showed that from 2000 to 2017, global R&D expenditure had expanded threefold' rising from, $722 billion to $2.2 trillion with the U.S. spending $549 billion and China spending $496 billion. However, in this same period, China's R&D spending has risen by an average of 17 percent per year, while the U.S.'s has been 4.3 percent. Consequently, China's spending on R&D may have already surpassed the U.S.

With the U.S. closing its technological doors, China will be even more committed to mastering and developing new technology. Indeed, China's 14th Five-Year Plan (2021-2025) is dedicated to becoming technologically self-reliant in semiconductor chips, artificial intelligence (AI), fifth-generation wireless technology and autonomous vehicles.

China has breezed past the global competition in 5G technology and it is already well into testing 6G technology.

On November 6, 2020, China successfully sent 13 satellites into orbit from Taiyuan Satellite Launch Center in Shanxi Province with one of them being the world's first 6G communications test satellite, which will be used to verify the performance of 6G technology in space.

The 6G frequency band will expand from the 5G millimeter wave frequency to the terahertz frequency which is an electromagnetic wave with a frequency range between microwave and infrared. The advantages of 6G over 5G are significant. Its speed will be over 100 times faster than 5G and its lossless transmission in space will enable communication across vast distances.

The University of Oulu in Finland predicts that 6G will lead to a future where physical to cyber fusion means augmented projection interfaces allow humans to project screens onto surfaces. In addition, physical and health data, currently only knowable with expensive hospital machines, will be actively monitored by individuals, in real time, who will have intimate knowledge of their biological data. Hardware devices as we currently know them will disappear. Instead, any surface can be transformed, on the fly, into a control surface and any space will become a holographic screen when desired. AI will be fused into the fabric of the world where all materials, even clothes, will have sensing abilities which communicate with one another. Such a world would be one where automation becomes the norm as machines, buildings and things communicate with one another. Energy consumption will be far more economical as smart power grids work in synchronicity to use energy efficiently.

The advanced automation that 6G brings to society will decrease the need for physical labor. For example, the automation of transport and consumption will decrease those working in these industries. This will also be the case for much white-collar work too, as smart systems know who has purchased what and "what is where", which will reduce the need for professions such as accountants and clerks. The ability to sense the human biological system has the possibility to revolutionize preventative medical treatment and even challenge the nature of work in the medical profession.

The 6G revolution will usher in a far more powerful means of production than we have today. As such, a 6G epoch will produce more surplus labor than we have currently. This surplus labor, with wise governance, can be transformed into the human defining labor of innovation and creativity which is inherent within all of us no matter what civilization we arise from.

C. Presentation

After you have read the article, please choose one of the following topics to develop your ideas. Make a presentation with PowerPoint to the class.

Topic 1   China's Advancements in 5G and 6G Technology
- Describe China's achievements in 5G technology and its ongoing efforts in testing and developing 6G technology.
- Highlight the significance of China's launch of the world's first 6G communications test satellite and its implications for the future of telecommunications.

Topic 2   Key Features and Advantages of 6G Technology
- Explain the key features and advantages of 6G technology, including its higher speed, expanded frequency band, and potential applications in space communication.
- Discuss the predictions made by the University of Oulu regarding the transformative effects of 6G technology on society.

Note:

- Craft a visually appealing PowerPoint with appropriate colors and images.
- Keep each slide concise, using fewer than 10 words.
- Use Pictures, illustrations or forms to make your point.
- Emphasize positive concepts and messages throughout your presentation.
- Aim to deliver your presentation smoothly, without relying on notes, within a timeframe of 4–5 minutes.

Video

A. Before You Watch

Read out the words below. Choose a word in the box to form an appropriate expression.

fiction   operator   decentralized   autonomous   tap   latency

- _____ network
- _____ into
- network _____
- lower _____
- science _____
- _____ cars

B. While You Watch

Exercise 1: Discussion

Do you think the possibilities and scenarios mentioned in the video can be realized?

Exercise 2: Dictation

Fill in the blanks with words and expressions you have heard from the video.

One expert claimed 6G could deliver mind boggling speeds of (1) _____ per second or 8000 gigabits per second. Forget using 5G to download just one movie in a few seconds from Netflix. With 6G speeds in just one second, you could download the entire Stranger Things series seven times over, or download (2)_____ of Netflix movies. It's great to talk about what 6G could do, but since it's just a pie in the sky idea at this point. Why does it matter to you? Well, it's going to be like 5G but more so even higher speeds, (3)_____ and masses of bandwidth.

Researchers and scientists are even talking about 6G being the network to move away from wires and use our devices as antennas to create a (4) _____ that's not under the control of a single (5)_____. So 6G could be the generation to take the power away from the big telecom companies and give it to the people. Five years expected to take tech that already exists, like (6) _____, drones and (7) _____, to the next level. But 6G may bring to life futuristic ideas like the integration of our (8) _____. Technology can soon tap into our bodies through something like (9) _____ and show a different world than reality. Imagine that 6G literally could bring on real life cyborgs. It can make charging your phone out of thin air a reality. And 6G coverage could extend over oceans and even into space.

So when we finally settle on the moon, you can still FaceTime with your boring earth friends. In other words, with 6G, science fiction could become (10) _____. 5G is going to change the way we communicate over the next five years, but when 6G is finally ready, it could entirely change the way we live.

Summary and Reflection

Now you have completed the chapter of 6G, it's time to reflect on your learning and ensure you have met the goals set for the chapter. Follow these steps to complete the checklist:

- Carefully read through the checklist provided, which outlines the key learning objectives and goals of the 6G chapter. For each item on the checklist, evaluate your own understanding and progress by checking the corresponding box.
- If you feel confident in your understanding and achievement of the goal, check the box; If you believe there are areas where you need further improvement or clarification, leave the box unchecked.

1. **Understanding of 6G Technology:**
   ☐ Can I explain what 6G technology is and its potential significance in telecommunications and beyond?
2. **Critical Thinking Skills:**
   ☐ Have I engaged in critical analysis of scenarios and predictions related to 6G technology?
   ☐ Have I evaluated the feasibility and implications of 6G technology in various contexts?
3. **Reading Comprehension:**
   ☐ Did I effectively skim and scan texts to extract relevant information about 6G technology and its impacts?
   ☐ Can I summarize and categorize information from complex texts using techniques such as multiple choice, mind mapping, and matching exercises?
4. **Vocabulary and Language Proficiency:**
   ☐ Have I expanded my vocabulary related to 6G technology and its terminology?
   ☐ Can I use technical terms accurately in an academic context?
5. **Collocating Words and Phrases:**
   ☐ Have I practiced using words and phrases related to 6G technology in appropriate contexts?
   ☐ Can I accurately collocate words and phrases to convey meaning effectively?
6. **Understanding of Impacts:**
   ☐ Have I explored the potential impacts of 6G technology on various industries and aspects of society?
   ☐ Can I articulate the economic, social, and technological implications of 6G technology?
7. **Presentation Skills:**
   ☐ Did I prepare and deliver a PowerPoint presentation on 6G technology within the specified time frame?
   ☐ Did I design my presentation with suitable colors, images, and concise text?
   ☐ Did I express positive ideas and concepts throughout my presentation?
8. **Reflection and Critical Analysis:**
   ☐ Have I reflected on the ideas and concepts learned throughout the chapter?
   ☐ Can I critically analyze the potential benefits and challenges associated with 6G technology?

Chapter 2
Chip

Objectives

In this chapter, you should be able to:
- Understand the fundamentals of semiconductor technology, including the structure and function of transistors, as well as advancements such as FinFET and nanosheet technologies.
- Develop critical thinking skills through tasks such as skimming, scanning, and matching to extract key information from technical texts related to chip technology.
- Improve reading comprehension, expand vocabulary, and enhance language proficiency by engaging with technical texts and terminology specific to semiconductor technology.
- Identify and analyze the collocation of words and phrases within the context of chip technology to deepen understanding of technical terminology and improve language usage.
- Enhance presentation skills by creating and delivering presentations on topics related to chip technology, utilizing tools like PowerPoint to organize information effectively and engage the audience.
- Reflect on and critically analyze knowledge acquired throughout the chapter, particularly in relation to potential future applications and impacts of chip technology advancements, considering different perspectives and evaluating significance.

Before You Read

A. Discussion
Look at the pictures below and discuss with a partner.
1. Below are the pictures of two transistors. How many parts are they made of?
2. What are the differences between these two pictures?
3. Read the descriptions of the transistors, and decide which one in the pictures is FinFET technology and which one is nanosheet technology.

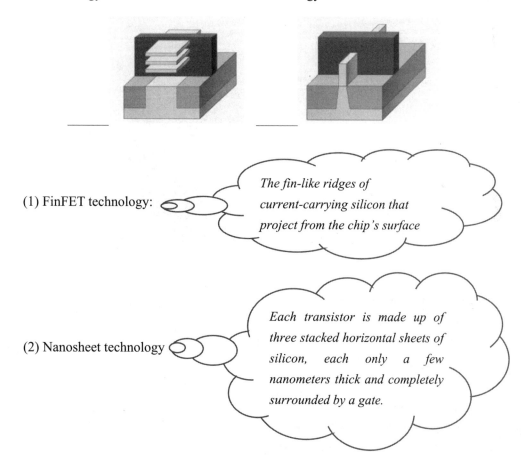

(1) FinFET technology: *The fin-like ridges of current-carrying silicon that project from the chip's surface*

(2) Nanosheet technology *Each transistor is made up of three stacked horizontal sheets of silicon, each only a few nanometers thick and completely surrounded by a gate.*

4. Name the different parts of the transistor below. Use the Internet or other resources to help you. You need to list both English and Chinese names of the components.

(1) _____

(2) _____

(3) _____

(4) _____

(5) _____

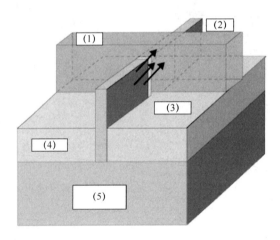

B. Skimming and Scanning

Browse the text and answer the questions below.

1. In what way does IBM outdo Samsung and TSMC?
A. By using 7-nm based chips.
B. By introducing 2-nm Node Chip.
C. By using FinFET technology.
D. By introducing three-sided gates.

2. What are the breakthrough developments in enabling 2-nm node chips?
A. The use of BDI to produce 12-nm gate lengths.
B. The application of EUV patterning to the FEOL.

C. The solutions to the interconnect issues.

D. The development of MTV scheme for SoC and HPC applications.

3. Which of the following statements are TRUE about 2-nm node chip?

A. It could reduce the carbon footprint of data centers.

B. It will prolong the cell phone battery life.

C. It will provide faster object detection for autonomous cars.

D. It will require users to charge their devices every night.

 Text

IBM Introduces the World's First 2-nm Node Chip
New Chip Milestone Offers Greater Efficiency and Performance

1. IBM has become the first in the world to introduce a 2-nm node chip. IBM claims this new chip will improve performance by 45 percent using the same amount of power, or use 75 percent less energy while maintaining the same performance level, as today's 7-nm-based chips. To give some sense of scale, with 2-nm technology, IBM could put 50 billion transistors onto a chip the size of a fingernail.

2. The foundation of the chip is nanosheet technology in which each transistor is made up of three stacked horizontal sheets of silicon, each only a few nanometers thick and completely surrounded by a gate. Nanosheet technology is poised to replace so-called FinFET technology named for the fin-like ridges of current-carrying silicon that project from the chip's surface. The life expectancy of FinFET has been more or less set at the 7-nm node. If it were to go any smaller, transistors would become difficult to switch off: Electrons would leak out, even with the three-sided gates.

3. One can't help but sense a bit of one-upmanship in IBM's development after Taiwan Semiconductor Manufacturing Co. (TSMC) decided to stay with FinFETs for its next generation process, the 3-nm node. While IBM's manufacturing partner, Samsung, does plan to use nanosheet technology for its 3-nm node chips, IBM outdid them both by using nanosheets and going down another step to a 2-nm node.

4. To further enable the chip beyond nanosheets, IBM has used bottom dielectric isolation (BDI) to produce 12-nm gate lengths, a feature representing a first in the industry. BDI involves the introduction of a dielectric layer underneath both the source and drain gate regions. The benefits of implementing a full BDI scheme are to reduce sub-channel leakage, to improve immunity to process variation, and to achieve power-performance improvement.

5. Another first for these chips was IBM's application of Extreme-Ultra-Violet(EUV) lithography patterning to the Front-End-of-Line(FEOL) where the individual devices (transistors, capacitors, resistors, etc.) are patterned in the semiconductor. After a decade of hand-wringing over whether EUV would ever deliver on its promises, it has in the last few years become a keystone for enabling 7-nm chips. Now, in this latest step in its evolution, EUV patterning has made it possible for IBM to produce variable nanosheet widths from 15-nm to 70-nm.

6. IBM has also developed a Multi-Threshold-Voltage(MTV) scheme for both System-on-a-Chip(SoC) and High-Performance Computing (HPC) applications. Threshold voltages—also known as gate voltages—are the minimum voltage differential needed between a gate and the source to create a conducting path between the source and drain terminals. MTV schemes leverage gates with different thresholds to optimize for power, timing, and area constraints.

7. While these all represent breakthrough developments in enabling 2-nm node chips, it does raise the question of interconnect crowding. In a press conference this week, Mukesh Khare, vice president of Hybrid Cloud at IBM Research in Albany, NY, addressed this question by explaining that this latest announcement is focused primarily on the transistor. According to Khare, the transistor is critical to address questions of scale, especially in scaling the gate length and the power and performance. However, he was quick to acknowledge the importance of interconnect issues.

8. "Interconnect scaling is equally important as the transistor," said Khare. "We are continuing to drive the correct scaling for the interconnects as well. That's part of our full 2-nm technology features."

9. Khare was reticent to discuss the specifics of standard cell library density and SRAM, and

only offered that it will likely follow the same bench marking that the industry has been tracking with 7-nm, 5-nm to 2-nm nodes.

10. IBM expects this chip design will be the foundation for future systems for both IBM and nonIBM chip players, and the potential benefits of these advanced 2-nm chips will be exponential for today's most advanced semiconductors.

11. The company anticipates that the 2-nm node could potentially reduce the carbon footprint of data centers. It estimates that if every data center changed their servers to 2-nm-based processors, it could save enough energy to power 43 million homes.

12. Closer to most of us is what IBM expects this to do our laptops and portable devices' functions—including quicker processing in applications, easier language translation, and faster 5G or 6G connections.

13. For those who find daily phone charging annoying, 2-nm node chips will quadruple cell phone battery life vs. 7-nm node chips, which the company says could require users to charge their devices only every third or fourth day, rather than every night.

14. IBM also anticipates that this may impact autonomous cars by providing faster object detection and reaction.

15. All of this sounds promising and it may not be that far off. Khare suggested that 2-nm chip modes could be rolling out of fabs as early as 2024.

 ## Reading Comprehension

A. Multiple Choice
Choose the best answer for each question.

1. What is the primary advantage of IBM's 2-nm node chip compared to current 7-nm-based chips?
   A. 45% improvement in performance with the same power consumption
   B. 75% reduction in energy consumption while maintaining the same performance level

C. Both A and B

D. None of the above

2. Which technology is poised to replace FinFET technology in IBM's 2-nm node chip?

A. Nanosheet technology

B. Extreme-Ultraviolet Lithography (EUV) Patterning

C. Bottom Dielectric Isolation (BDI)

D. Multi-Threshold-Voltage (MTV) Scheme

3. What is the purpose of Extreme-Ultraviolet Lithography (EUV) Patterning in IBM's chip manufacturing process?

A. To enhance transistor performance

B. To reduce sub-channel leakage

C. To facilitate the production of variable nanosheet widths

D. All of the above

4. According to IBM, what is the potential environmental benefit of adopting 2-nm-based processors in data centers?

A. Reduction in carbon footprint equivalent to powering 43 million homes

B. Reduction in water consumption by 50%

C. Increase in energy consumption by 20%

D. No environmental impact

5. How does IBM expect the 2-nm node chip to impact the functionality of laptops and portable devices?

A. Slower processing speed

B. Difficulties in language translation

C. Faster processing in applications

D. Decreased battery life

B. Mind Map

How many main parts do you think the article is composed of? Group the paragraphs and fill in the blanks with the information you read from the article.

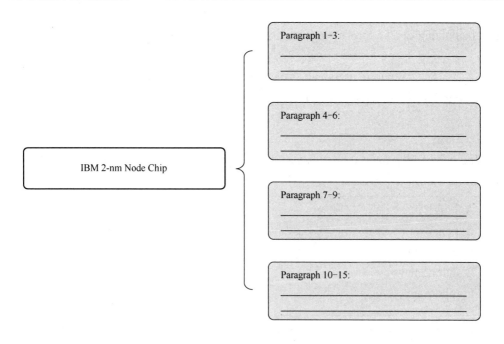

C. Matching

Read the text and decide which paragraph mentions the following information? Write the number of the paragraph before each sentence.

_____ 1. This technological advancement is underpinned by nanosheet architecture, which replaces the traditional FinFET technology used in current chips.

_____ 2. To address challenges associated with transistor miniaturization, IBM has introduced Bottom Dielectric Isolation(BDI), a novel technique enhancing gate length and power-performance characteristics.

_____ 3. IBM's application of Extreme-Ultra-Violet(EUV) lithography patterning enables the production of nanosheets with variable widths, marking a pivotal advancement in chip manufacturing.

_____ 4. Multi-Threshold-Voltage(MTV) schemes have been developed to optimize power, timing, and area constraints for both System-on-a-Chip(SoC) and High-Performance Computing(HPC) applications.

_____ 5. While transistor development takes center stage, IBM acknowledges the importance of addressing interconnect crowding issues in chip design.

_____ 6. The environmental benefits of adopting 2-nm-based processors in data centers are substantial, potentially reducing carbon footprints equivalent to powering millions of homes.

_____ 7. IBM foresees significant improvements in consumer electronics, with 2-nm chips facilitating quicker processing, language translation, and faster connectivity such as 5G or 6G.

_____ 8. Cell phone users can expect a quadrupled battery life with 2-nm node chips compared to 7-nm counterparts, potentially reducing the need for frequent charging.

D. Cloze

The information below is a summary of the text. Complete the summary by filling in the blanks with the words provided.

A. FinFET	F. battery
B. transistor	G. milestone
C. efficiency	H. nanosheet
D. autonomous	I. isolation
E. lithography	J. carbon

IBM has introduced the world's first 2-nm node chip, marking a significant (1) _____ in semiconductor technology. This breakthrough promises notable enhancements in both (2) _____ and performance. The chip's foundation lies in (3) _____ technology, a novel architecture poised to replace the current (4) _____ technology. IBM has also introduced innovative techniques such as bottom dielectric _____ (BDI) and Extreme-Ultra-Violet(EUV) (5) _____ patterning to address challenges associated with (6) _____ miniaturization and improve chip manufacturing. Multi-Threshold-Voltage(MTV) schemes have been developed to optimize (7) _____ and timing constraints for various applications. While the focus remains on transistor development, IBM acknowledges the importance of addressing power crowding issues in chip design. The environmental benefits of adopting 2-nm-based processors in data centers are substantial, potentially reducing (8) _____ footprints equivalent to powering millions of homes. Moreover, consumer electronics are expected to experience significant improvements, including quicker processing, language translation, and faster connectivity. Cell phone users can anticipate a quadrupled (9) _____ life compared to previous generations of chips. Beyond consumer electronics, IBM foresees benefits for (10) _____ vehicles, such as faster object detection and reaction times.

Language Building

A. Glossary

Proper Nouns
IBM 　　国际商业机器公司，又称万国商业机器公司，总公司在纽约州阿蒙克市。该公司于 1911 年由托马斯·沃森创立于美国，是全球最大的信息技术和业务解决方案公司。 **FinFET technology** 　　FinFET 全称 Fin Field-Effect Transistor，中文名叫鳍式场效应晶体管，是一种新的互补式金氧半导体晶体管。FinFET 因晶体管的形状与鱼鳍相似而得名。这种设计可以改善电路控制并减少漏电流，缩短晶体管的闸长。 **TSMC** 　　台湾积体电路制造股份有限公司，简称台积电，属于半导体制造公司。该公司成立于 1987 年，是全球第一家专业集成电路制造服务（晶圆代工）企业，总部与主要工厂位于中国台湾省的新竹市科学园区。 **Bottom Dielectric Isolation (BDI)** 　　底部介电层，能够有效减少漏电流，降低芯片功耗。 **Extreme-Ultra-Violet(EUV) Lithography** 　　极紫外辐射，是波长在 124nm 到 10nm 之间的电磁辐射，对应光子能量为 10eV 到 124eV。人工 EUV 可由等离子源和同步辐射源得到，其主要用途包括光电子谱，对日 EUV 成像望远镜，光微影等。 **Front-End-of-Line (FEOL)** 　　前道工艺，是整个芯片生产过程中关键的环节之一，涉及芯片的前期准备、光刻、刻蚀、薄膜沉积、离子注入、清洗、CMP、量测等工艺；这些环节一旦出现问题，会对整个芯片生产流程造成影响。

Hybrid Cloud
混合云，融合了公有云和私有云，是近年来云计算的主要模式和发展方向。

Static Random-Access Memory(SRAM)
静态随机存取存储器，是随机存取存储器的一种。所谓的"静态"，是指这种存储器只要保持通电，里面储存的数据就可以恒久保持。

Academic Words	
maintain(v.)	包含，含有
transistor(n.)	晶体管
fingernail(n.)	手指甲
stack(n.)	垛，堆
horizontal(a.)	水平的
silicon(n.)	硅
leakage(n.)	泄露，透露
dielectric(a.)	非传导性的（绝缘的）
scheme(n.)	计划，方案
differential(n.)	差值
terminal(n.)	终端
interconnect(v.)	互相联系
feature(n.)	特点，特征
reticent(a.)	沉默的，不愿说的
exponential(a.)	越来越快的；指数的
estimate(v.)	估计；判断
autonomous(a.)	自动的

B. Words and Phrases

Exercises 1   Word Choice
Use the words in the box to finish the sentences.

transistor	reticent	dielectric	scheme	drain
threshold	exponentially	autonomous	constraint	leak

1. The smaller the chip, the more likely electrons are to_____ away from the source due to quantum tunneling.

2. A spokesman for the company was _____ to disclose the exact technical details of the

project.

3. With the rapid development of integrated circuit technology, a chip can contain more_____.

4. The wide use of the Internet has made people more connected, and the number of families with computers has increased _____.

5. The boss ordered the group of employees to come up with a _____ to defeat their competitors within a week.

Exercise 2   Phrases
Match the words provided below with appropriate one in the box.

1. _____ layer	nanosheet
2. _____ device	portable
3. threshold _____	voltage
4. _____ technology	footprint
5. carbon _____	dielectric

Exercise 3   Sentence Completion
Complete the sentences by filling in the blanks with phrases in the above exercise.

1. _____ is an important parameter that should be paid attention to in the application of devices.

2. Making good use of _____ can greatly improve office efficiency.

3. _____ is one of the most popular semiconductor technologies in recent years.

4. _____ can effectively isolate heat, prevent overheating of the wire caused by equipment damage.

5. _____ is historically defined as the total set of Greenhouse Gas (GHG) emissions caused by an organization, event, product or person

Exercise 4   Translation
Translate the sentences by using the words and phrases you have learned in the above two exercises.

1. 利用<u>纳米片技术</u>，有望使芯片的尺寸进一步缩小。

_____

2. 许多电子器件的性质在<u>阈值电压</u>两侧会存在显著的差别。

_____

3. 近年来，许多便携设备的出现使得人们的工作和生活更加便利。
_____

4. 这家环保协会呼吁人们少开私家车，多骑单车出行，以减少碳足迹。
_____

5. 在发生事故时，一个良好的介电层有助于保护用电人员的安全。
_____

C. Collocation

Exercise 1   Modifiers

Find out the adjectives that modify the noun "performance" in the article. The first letter has been provided.

i_____   performance(para.1)
s_____   performance(para.1)
p_____   performance(para.4)
h_____   performance(para.6)

Exercise 2   Blank Filling

Scan the text and complete the sentences containing the word "performance".

1. IBM claims this new chip will _____ performance by 45 percent using the same amount of power.

2. This new chip will use 75 percent less energy while _____ the same performance level, as today's 7-nm-based chips.

3. This method is usually intended for environments where _____ performance is more important than reliability.

4. These are irrelevant and don't _____ the performance of the application.

5. The _____ performance, fuel economy, and exhaust emission of the vehicle engines are improved considerably through the electronic control.

6. A living example is introduced to demonstrate how to _____ the performance problem of Java software with the aid of the tool developed.

7. If the simulation results are not matched by the _____ performance, some work needs to be done.

8. These new materials possibly reduce the cost and improve _____ performance.

9. Linux _____ performance was not degraded during the long duration of the run.

10. Energy efficiency is one of the critical parameters to _____ the performance of wireless networks.

Exercise 3　Translation

Translate the sentences below from Chinese to English using "performance" and its collocation in this section.

1. 我们设计了一组实验来评估混合模型的性能。
_____

2. 提取主要的功能码以达到简化协议、增强控制器性能的目的。
_____

3. 通过利用这些特性，您可以更有效地管理数据，同时提供最佳的数据库性能。
_____

4. 我们需要解决经常出现的质量问题，这些问题会威胁到企业经营业绩。
_____

5. 该系统将在客户信息记录、售后服务支持和销售业绩分析等方面显示出其强大的功能。
_____

D. Terminology

Exercise 1　Table Filling

Read the article and find the English technical terms according to the Chinese equivalents.

English Technical Terms	Chinese Equivalents
	纳米片技术
	底部介电隔离
	高性能计算
	阈值电压
	混合云
	便携式设备

Exercises 2　Blank Filling

Use the terms in the above exercise to complete the sentences below.

1. The foundation of the chip is _____ in which each transistor is made up of three stacked horizontal sheets of silicon, each only a few nanometers thick and completely surrounded by a gate.

2. To further enable the chip beyond nanosheets, IBM has used _____ to produce 12-nm gate lengths, a feature representing a first in the industry.

3. IBM has also developed a Multi-Threshold-Voltage (Multi-VT) scheme for both System-on-a Chip (SoC) and _____ applications.

4. _____ —also known as gate voltages—are the minimum voltage differential needed between a gate and the source to create a conducting path between the source and drain terminals.

5. In a press conference this week, Mukesh Khare, vice president of _____ at IBM Research in Albany, NY, addressed this question by explaining that this latest announcement is focused primarily on the transistor.

6. Closer to most of us is what IBM expects this to do our laptops and _____ functions—including quicker processing in applications, easier language translation, and faster 5G or 6G connections.

Critical Reading and Writing

A. Brainstorming

Work in groups. Fill in the table according to the instruction.

Future Applications Exploration: Choose one potential future application of IBM's 2-nm node chip mentioned in the article (e.g., data centers, consumer electronics, autonomous vehicles). Each group brainstorms and discusses possible specific uses and benefits of the chip in their assigned application area.

Future Applications Exploration

1. Data Centers:
2. Consumer Electronics:

3. Autonomous Vehicles:

B. Critical Reading

Exercise 1: The following are the answers to the question above provided by ChatGPT. Read each item carefully and categorize them into the three dimensions according to their applications.

Data Centers	
Consumer Electronics	
Autonomous Vehicles	

⑤ ChatGPT

a. Enhanced energy efficiency may lead to extended battery life for portable devices, reducing the need for frequent charging and improving user convenience.

b. Enhanced data processing capabilities could enable faster analysis of large datasets, leading to improved decision-making and insights for businesses.

c. Improved energy efficiency may contribute to extending the range and battery life of electric autonomous vehicles, enhancing their practicality and reducing operational costs.

d. Smartphones, tablets, and laptops equipped with the 2-nm node chip could experience significantly faster processing speeds, resulting in smoother multitasking and quicker app launches.

e. The high processing power of the 2-nm node chip could enhance the performance of autonomous vehicles, enabling faster and more accurate object detection, recognition, and decision-making.

f. Increased processing power and energy efficiency could allow data centers to handle larger workloads while reducing operational costs.

Exercise 2: Can you give more examples as what ChatGPT has provided? List as least 2 items for each category. Use the Internet to assist you if necessary.

C. Essay Writing

Having engaged in discussions and vocabulary preparation, you've likely generated numerous insightful ideas. Now, it's time to reflect on these ideas and the knowledge you've acquired by crafting an essay. Use the following instructions to guide your writing process:

Topic

Exploring the Potential Applications of IBM's 2-nm Node Chip in ... Industry.

Background Information

This chip represents a significant leap forward in efficiency and performance, with potential applications ranging from data centers to consumer electronics and autonomous vehicles. Imagine you are a technology analyst tasked with exploring the implications of IBM's recent introduction of the world's first 2-nm node chip. In a well-structured essay, critically analyze the significance of this technological advancement and its potential impact on various aspects of industries, such as data centers, consumer electronics, and autonomous vehicles.

Instructions

In an essay of approximately 300–450 words, analyze the potential applications of IBM's 2-nm node chip in a specific industry. Provide specific evidence and arguments to support your opinion.

Your essay should include the following components:

Introduction (approximately 50–75 words): Briefly introduce the topic and provide context for your analysis.

- State your thesis or main argument regarding the industry you will be focusing on and highlight the significance of IBM's 2-nm node chip in revolutionizing technological capabilities within that industry.

Body Paragraphs (approximately 250–350 words): Present your argument in detail, supported by specific evidence and examples in 3 separate paragraphs.

- Present your analysis of the potential applications of IBM's 2-nm node chip in the chosen industry in detail.
- Discuss specific areas where the chip's advanced capabilities could be leveraged to drive innovation, improve efficiency, and create new opportunities.
- Consider factors such as performance enhancements, energy efficiency, and scalability.

- Provide evidence and examples to support your arguments, drawing from the features and benefits of IBM's 2-nm node chip as described in the text.

Conclusion (approximately 50–75 words): Summarize your main points and restate your thesis in light of the evidence presented.
- Reflect on the transformative impact that the chip could have on operations, business models, and technological advancements within the industry.
- Consider the broader implications of adopting this advanced chip technology and potential avenues for future research and exploration.

* Ensure that your essay is well-structured, logically organized, and supported by evidence from reputable sources. Use clear and concise language, and proofread your work carefully for grammar, punctuation, and spelling errors.

Discussion and Presentation

A. Group Discussion
Exercise 1: Think and discuss the questions below.

1. How do you think the ban on exports of advanced GPU chips to China will impact the global semiconductor industry in the long term? Consider factors such as technological innovation, supply chain disruptions, and geopolitical tensions.

2. What are the potential implications of China's response to the U.S. chip ban, particularly in terms of its efforts to achieve technological self-reliance and sustainability in the semiconductor industry? Discuss the challenges and opportunities for domestic players in China.

3. How might the U.S. chip ban influence the strategies and business models of semiconductor companies, both in China and globally? Consider the implications for research and development, investment, and collaboration efforts within the semiconductor ecosystem.

Exercise 2: Read the following article. Search for any information related to the above questions. Do you think the answers provided in the article match yours? Share your ideas with the class.

B. Extended Reading-Chips in China

Read the article, highlight any ideas or language that will help you in your presentation in the next session.

Experts: What Will China Win Out of U.S. Latest GPU Chip Ban?

The U.S. has once again imposed a ban on exporting chips to China, specifically targeting sophisticated Graphics Processing Units (GPUs). This move is believed to be aimed at further limiting China's technological capabilities. In response, China has expressed firm opposition to what it perceives as hegemonic behavior. Despite this, China remains optimistic about its growing high-tech expertise and applications in the field. Chip designer NVIDIA Corp recently disclosed that U.S. officials instructed them to cease exporting two leading computing chips for AI work to China. NVIDIA reported that the U.S. government issued new license requirements for future exports to China, affecting its A100 and forthcoming H100 integrated circuits. However, the U.S. subsequently permitted certain exports and technology transfers necessary for the development of the H100 chip. Additionally, Advanced Micro Devices (AMD) is reported to have received similar license requirements affecting its MI250 AI chips, although the company has not confirmed this publicly.

During a routine press briefing, Chinese Foreign Ministry spokesperson Wang Wenbin criticized the U.S. decision, labeling it as a typical example of "sci-tech hegemony." He argued that the U.S., leveraging its technological advantages, has misused the concept of national security to stifle the development of emerging economies and developing countries. Wang asserted that such actions violate market economy principles, disrupt international economic and trade orders, and destabilize global industrial and supply chains, which China firmly opposes. Furthermore, Wang accused the U.S. of politicizing and weaponizing science, technology, and business issues, aiming to monopolize advanced technologies and maintain its hegemony. He emphasized that these efforts to push for decoupling will not succeed. Similarly, Chinese Commerce Ministry spokesperson Shu Jueting expressed concerns about the U.S. move, highlighting its adverse effects on the legitimate rights of Chinese firms and the interests of the U.S. firms. Shu urged the U.S. to rectify its actions promptly and treat companies from all countries, including Chinese companies, fairly to promote global economic stability.

In the era of 5G technology, the global supply chain has remained stable until recent disruptions caused by the U.S. bans on chip exports to China. These bans specifically targeted

Graphics Processing Units (GPUs), essential for various computing applications, including Artificial Intelligence (AI) and creative production. China, being the largest market for chip products, witnessed significant growth in semiconductor equipment sales, reaching $29.62 billion in 2021. However, experts view the U.S. ban as part of a long-term strategy to curb China's technological advancement and maintain dominance. While some analysts believe that the impact of the ban on China's domestic companies will be limited, others foresee broader implications, especially concerning advanced computing power essential for AI and supercomputing. Additionally, the U.S. crackdown extends beyond chip exports, encompassing vital technologies like Extreme Ultra-Violet(EUV) lithography systems. Amid these challenges, Chinese firms are intensifying efforts to enhance domestic capabilities in chip manufacturing, with startups venturing into research and development. Despite the dominance of companies like NVIDIA, there is a growing push for self-reliance in GPU production and the development of core technologies within China's chip industry. This shift towards domestic innovation underscores the necessity for increased investment and talent attraction to ensure long-term sustainability and technological autonomy.

C. Presentation

After you have read the article, please choose one of the following topics to develop your ideas. Make a presentation with PowerPoint to the class.

Topic 1 Understanding the Implications of the U.S. GPU Chip Ban on China's Technological Ambitions
- Analyze the potential impact of the ban on China's technological capabilities, particularly in the field of Artificial Intelligence (AI) and creative production.
- Discuss the reactions and responses from Chinese government officials and industry stakeholders to the ban.
- Explore the broader geopolitical implications of the ban on global supply chains and technological competition between the U.S. and China.
- Provide insights into potential strategies for China to navigate and mitigate the effects of the ban on its technological ambitions.

> **Topic 2 Charting China's Tech Trajectory: Exploring Strategies for Self-Reliance in Response to the U.S. Chip Export Restrictions**
> - Examine the initiatives and investments undertaken by Chinese firms and startups to enhance domestic capabilities in chip manufacturing and innovation.
> - Highlight specific examples of Chinese companies venturing into research and development to reduce reliance on imported chips.
> - Discuss the challenges and opportunities faced by China in building a sustainable and competitive semiconductor industry.

Note:
- Craft a visually appealing PowerPoint with appropriate colors and images.
- Keep each slide concise, using fewer than 10 words.
- Use Pictures, illustrations or forms to make your point.
- Emphasize positive concepts and messages throughout your presentation.
- Aim to deliver your presentation smoothly, without relying on notes, within a timeframe of 4–5 minutes.

 Video

Video

A. Before You Watch

Read out the words below. Choose a word in the box to form an appropriate expression.

scaling	design	leaking	capabilities	computing

- nanosheet_____
- _____technology
- _____up
- _____through
- manufacturing _____

B. While You Watch

Exercise 1: Discussion

Watch a video about IBM's breakthrough developments in enabling 2-nm node chips. Why the size of a chip matter so much? Explain in detail.

Exercise 2:
Fill in the blanks with words and expressions you have heard from the video.

This breakthrough invention allows intel to create (1)_____that are smaller, faster, and use less power than ever before. Enabling a new generation of (2)_____in every category. This new design turns the channel from that FIN into a stack of tiny ribbons or nanosheets with the gate material (3) _____, entirely around them.

In fact, while people do seem to be calling this (4) _____. Now, when I first heard about it, it was being referred to as gate all around. As transistors get tinier and tinier, this problem of keeping current from (5)_____has been one of the main challenges. And this nanosheet design is being claimed as a (6)_____. These designs are more difficult to make, and they require many astronomically thin layers of material stacked and built into complex structures all at the nanometer scale.

But they do actually offer some big potential advantages, including (7)_____in different parts of the chip to accommodate different electrical needs, or even a proposed intel design that just stacks multiple layers of transistors on top of each other. IBM's manufacturing partner on these new chips is Samsung, and they've even announced their exploring nano sheet design for their upcoming through (8)_____. Samsung has also been aggressively scaling up their EUV or extreme ultraviolet (9)_____. And you can bet these nanosheet designs will almost certainly require that technology to achieve these tiny sizes. Interestingly, TSMC the chipmaker who currently (10)_____crown with their cutting edge price projects has said they're actually sticking with thin fed design, at least through their 3-nm node.

📖 Summary and Reflection

Now that you have completed the chapter of chip, it's time to reflect on your learning and ensure you have met the goals set for the chapter. Follow these steps to complete the checklist:
- Carefully read through the checklist provided, which outlines the key learning objectives and goals of the chip chapter. For each item on the checklist, evaluate your

own understanding and progress by checking the corresponding box.
- If you feel confident in your understanding and achievement of the goal, check the box; If you believe there are areas where you need further improvement or clarification, leave the box unchecked.

1. **Understanding of Chip Technology:**
   ☐ Have I deepened my understanding of semiconductor technology, including the structure of transistors and advancements like FinFET and nanosheet technologies?
   ☐ Can I explain these concepts clearly to others?

2. **Critical Thinking Skills:**
   ☐ How effectively did I apply critical thinking skills to tasks such as skimming, scanning, and matching to extract key information from technical texts?
   ☐ Did I identify relevant information and draw logical conclusions?

3. **Reading Comprehension, Vocabulary, and Language Proficiency:**
   ☐ Have I improved my reading comprehension, vocabulary, and language proficiency through exercises focused on technical texts and terminology?
   ☐ Can I accurately define and use technical terms in context?

4. **Collocating Words and Phrases:**
   ☐ Can I identify and analyze the collocation of words and phrases within the context of chip technology to enhance my understanding of technical terminology?
   ☐ Have I practiced using these collocations in sentences?

5. **Understanding of Impacts:**
   ☐ How did I explore the geopolitical impacts of the U.S. ban on GPU chip exports to China, considering factors such as technological innovation, supply chain disruptions, and global economic stability?
   ☐ Can I articulate the broader implications of such actions?

6. **Presentation Skills:**
   ☐ How did I develop my presentation skills by creating and delivering presentations on topics related to chip technology, using tools like PowerPoint?
   ☐ Did I effectively organize information, use visual aids, and engage the audience?

7. **Reflection and Critical Analysis:**
   ☐ How did I reflect on and critically analyze the knowledge acquired throughout the chapter, particularly in relation to potential future applications and impacts of chip technology advancements?
   ☐ Did I consider different perspectives and evaluate the significance of these advancements?

Chapter 3
AI Image Generation Technology

> **Objectives**
> In this chapter, you should be able to:
> - Analyze and discuss the latest developments and potential applications of AI image generation technology in various fields such as entertainment, healthcare, and advertising.
> - Develop critical thinking, reading, and writing skills by evaluating the ethical, legal, and technical implications of AI-generated images in different contexts.
> - Demonstrate the ability to summarize and categorize information from complex texts related to AI image generation using techniques such as multiple choice, mind mapping, and matching exercises.
> - Expand vocabulary and language proficiency in an academic context through exercises like glossaries, word choice, and translation tasks related to AI image generation and its terminology.
> - Engage in collaborative discussions and presentations to share insights and perspectives on AI image generation technology and its impact on industries and society, both before and after reading relevant material.

 Before You Read

A. Discussion
Look at the pictures below and discuss with a partner.

1. Which pictures do you think are generated by AI? Why?

2. What prompts might be used to generate such pictures? Use your prompts in an AI image generator, and see if it can generate similar pictures.

B. Skimming and Scanning

Browse the text and answer the questions below.

1. What is LostGANs?

A. It is an AI model for image generation.

B. It is developed by two Ph.D. students from Australian university.

C. The paper on LostGANs is published in the journal of IEEE Transaction.

D. Compared with GANs, LostGANs can generate more consistent images when the input layout is reconfigurable.

2. To how much pixels of resolution can LostGANs synthesize images compared with prior AI modes?

A. 150 × 150

B. 1920 × 1080

C. 640 × 480

D. 512 × 512

3. For what scenarios can LostGANs be used to help generate images according to the passage?

A. Education

B. Online shopping

C. News broadcasting

D. Autonomous vehicle

Text

Harnessing the Wild Power of AI Image Generation
A Layout and Style-Based Architecture Shows How to Control AI Capabilities to Generate Complex Images

1. AI has already shown off the capability to create photorealistic images of cats, dogs, and people's faces that never existed before. More recently, researchers have been investigating how to train AI models to create more complex images that could include many different objects arranged in different poses and configurations.

2. The challenge involves figuring out how to get AI models—in this case typically a class of deep learning algorithms known as Generative Adversarial Networks(GANs)—to generate more controlled images based on certain conditions rather than simply spitting out any random image. A team at North Carolina State University has developed a way for GANs to create such conditional images more reliably by using reconfigurable image layouts as the starting point.

3. "We want a model that is flexible enough such that when the input layout is reconfigurable, then we can generate an image that can be consistent," says Tianfu Wu, an assistant professor in the department of electrical and computer engineering at North Carolina State University in Raleigh.

4. This layout and style-based architecture for GANs (nicknamed LostGANs) came out of research by both Wu and Wei Sun, a former Ph.D. student in the department of electrical and computer engineering at North Carolina State University who is currently a research scientist at Facebook. Their paper on this work was published last month in the journal IEEE Transactions on Pattern Analysis and Machine Intelligence.

5. The starting point for the LostGANs approach involves a simple reconfigurable layout that includes rectangular bounding boxes showing where a tree, road, bus, sky, or person should be within the overall image. Yet previous AI models have generally failed to create

photorealistic and perfectly proportioned images when they tried to work directly from such layouts.

6. This is why Wu and Sun trained their AI model to use the bounding boxes in the layout as a starting point to first create "object masks" that look like silhouettes of each object. This intermediate "layer-to-mask" step allows the model to further refine the general shape of such object silhouettes, which helps to make a more realistic and final "mask-to-image" result where all the visual details have been filled in.
7. The team's approach also enables researchers to have the AI change the visual appearance of specific objects within the overall image layout based on reconfigurable "style codes." For example, the AI can generate different versions of the same general wintry mountain landscape with people skiing by making specific style changes to the skiers' clothing or even their body pose.

8. The results from the LostGANs approach are still not exactly photorealistic—such AI-generated images can sometimes resemble impressionistic paintings with strangely distorted proportions and poses. But LostGANs can synthesize images at a resolution of up to 512 × 512 pixels compared to prior layout-to-image AI models that usually generated lower-resolution images. The LostGANs approach also demonstrated some performance improvements over the competition during benchmark testing with the COCO-Stuff dataset and Visual Genome dataset.

9. Next step for LostGANs could involve better capturing the details of interactions between people and small objects, such as a person holding a tennis racket in a certain way. One way that LostGANs might improve here would be to use "part-level masks" that represent various components making up an object.

10. But just as importantly, Wu and Sun showed how to train LostGANs more efficiently using fewer labeled conditions without having to sacrifice the quality of the final image. Such semi-supervised training can rely on just 50 percent of the usual training images to bring LostGANs up to its usual performance standards. The source code and pretrained models of LostGANs are available online at GitHub for any other researchers interested in giving this approach a try.

11. Tech companies and organizations with much deeper pockets than academic labs have already begun showing the potential of harnessing AI-generated images. In 2019, NVIDIA demonstrated an AI art application called GauGAN that can convert rough sketches drawn by human artists into realistic-looking final images. In early 2021, OpenAI showed off a DALL·E version of its GPT-3 language model that can convert text prompts such as "an armchair in the shape of an avocado" into a realistic final image.

12. Still, the LostGANs research has a lot to offer despite not yet achieving as polished image results. By taking the layout-to-mask-to-image approach, LostGANs enables researchers to better understand how the AI model is generating the various objects within an image. Such transparency offered by LostGANs represents an improvement on the typical "black box" approach to many AI models that can leave even experts scratching their heads over how the final image was generated.

13. "For example, if you look at the image and the person doesn't look correct, you can trace it back and see that it's because the mask is not correctly computed," Wu explains. "The mask is better for understanding what's going on in the generated image and also makes it easier to control the image generation."

14. The research could eventually help robots and AI agents to better envision the results of future interactions with objects within their immediate environment. Such image generation based on reconfigurable layouts could also potentially help generate different visual scenarios that could help train autonomous vehicles.

15. And in the near-term, LostGANs could play the role of an educational tool that invites students and other curious learners to interact with AI through setting up a simple image layout. During a departmental open house, an early version of LostGANs attracted the attention of local high school students with its still imperfect AI-generated images.

16. "I think that will be fun for those students to play with," Wu says. "Then they can get a rough understanding that 'Oh, this is something where I can interact with an AI system through this simple painting.'"

Reading Comprehension

A. Multiple Choice

Choose the best answer for each question.

1. What is the primary goal of the LostGANs approach developed by researchers at North Carolina State University?

   A. To generate simpler, single-object images

   B. To create complex images based on reconfigurable layouts

   C. To focus solely on improving the resolution of images

   D. To eliminate the need for style codes in image generation

2. What role do "reconfigurable image layouts" play in the LostGANs system?

   A. They are used to finalize the visual details in an image.

   B. They serve as the starting point for generating conditional images.

   C. They help in training the model without any human input.

   D. They replace the need for object masks in the generation process.

3. According to the article, how do LostGANs handle the initial phase of image generation?

   A. By directly converting text prompts into images

   B. By using object masks that resemble silhouettes of each object

   C. By creating a detailed color scheme for each object first

   D. By sketching rough outlines that are later refined

4. Which feature allows LostGANs to modify the appearance of specific objects within the overall image layout?

   A. High-resolution image processing

   B. Rectangular bounding boxes

   C. Reconfigurable style codes

   D. Layer-to-mask steps

5. What potential future applications of LostGANs are mentioned in the article?
A. Generating images for social media platforms
B. Assisting robots in envisioning interactions with objects
C. Creating educational tools for medical training
D. Enhancing video resolution for streaming services

B. Mind Map

How many main parts do you think the article is composed of? Group the paragraphs and fill in the blanks with the information you read from the article.

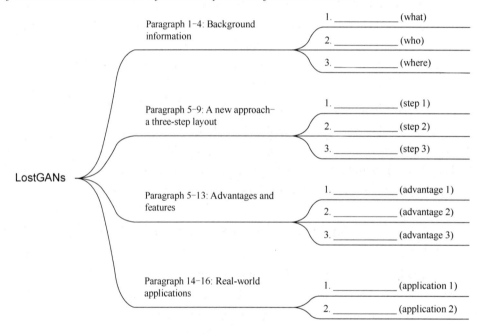

C. Matching

Read the text and decide which paragraph mentions the following information? Write the number of the paragraph before each sentence. The sentences are randomly ordered, which is a bit more challenging.

_____ 1. Unlike previous AI models, LostGANs start by creating object masks from the layout, refining the general shape of each object silhouette before generating the final image.

_____ 2. While the results from LostGANs may not always be photorealistic, they can synthesize images at higher resolutions compared to previous models.

_____ 3. Researchers at North Carolina State University have developed an innovative approach to AI image generation that focuses on creating complex images based on flexible layouts.

_____ 4. The approach taken by LostGANs enables researchers to better understand how AI models generate images by providing transparency in the generation process.

_____ 5. LostGANs also allow for modifications in the appearance of specific objects within the image layout using reconfigurable style codes, offering versatility in image generation.

_____ 6. LostGANs could potentially aid in envisioning future interactions between robots and objects in their environment, contributing to advancements in robotics.

_____ 7. Moreover, LostGANs could serve as educational tools, allowing students to interact with AI systems and understand the basics of image generation.

_____ 8. Overall, the research on LostGANs offers valuable insights into the capabilities and limitations of AI image generation, paving the way for future advancements in the field.

_____ 9. This new method, known as LostGANs, utilizes reconfigurable image layouts as a starting point for generating conditional images, aiming for more controlled results.

_____ 10. This transparency is beneficial for troubleshooting and refining the generated images, ensuring better accuracy and realism.

D. Cloze

The information below is a summary of the text. Complete the summary by filling in the blanks with the words and phrases provided.

A. envisioning future interactions	F. transparency
B. educational tools	G. complex images
C. object masks	H. advancements in robotics
D. photorealistic	I. limitations
E. innovative approach	J. higher resolutions

Researchers at North Carolina State University have developed an (1)_____ to AI image generation known as LostGANs. This method focuses on creating complex images based on (2)_____ and reconfigurable style codes. Unlike previous AI models, LostGANs start by creating (3)_____ from the layout, refining the general shape of each object silhouette before generating the final image. While the results from LostGANs may not always be (4)_____, they can synthesize images at (5)_____ compared to previous models. The (6)_____ provided by LostGANs in the generation process enables researchers to better understand how AI models generate images and refine them accordingly.

Moreover, LostGANs could potentially aid in (7)_____ between robots and objects in their environment, contributing to (8)_____ in robotics. Additionally, LostGANs could serve as (9)_____, allowing students to interact with AI systems and understand the basics of image generation. Overall, the research on LostGANs offers valuable insights into the capabilities and (10)_____ of AI image generation, paving the way for future advancements in the field.

 Language Building

A. Glossary

Proper Nouns
Generative Adversarial Network(GAN) 生成对抗网络 **North Carolina State University** 北卡罗来纳州立大学 **the COCO-Stuff dataset** COCO 数据集,其是一个大型的、丰富的物体检测、分割和字幕数据集。这个数据集以场景理解为目标,主要从复杂的日常场景中截取,图像中的目标通过精确的分割进行位置的标定。 **Visual Genome(VG) dataset** 分割 VG 数据集是斯坦福大学李飞飞组于 2016 年发布的大规模图片语义理解数据集,他们希望该数据集能像 ImageNet 那样推动图片高级语义理解方面的研究。 **GitHub** GitHub 是一个面向开源及私有软件项目的托管平台,因为只支持 git 作为唯一的版本库格式进行托管,故名 GitHub。

NVIDIA

英伟达是一家人工智能计算公司，创立于 1993 年，总部位于美国加利福尼亚州圣克拉拉市。美籍华人 Jensen Huang（黄仁勋）是其创始人兼 CEO。

GauGAN

GauGAN 是一款专门用于神经网络作图的软件。GauGAN 最新版能为从建筑师、城市规划者到景观设计师和游戏开发商提供一个强大的工具来创建虚拟世界。GauGAN 官方版使用户可以画出语义分割图，并合成自己需要的场景，其中，这些分割图都需要相应的标签，如天空、大海或者雪。

OpenAI

OpenAI 是由诸多硅谷、西雅图科技大亨联合建立的人工智能非营利性组织。2015 年，马斯克与其他硅谷、西雅图科技大亨进行连续对话后，决定共同创建 OpenAI，希望能够预防人工智能的灾难性影响，推动人工智能发挥积极作用。

Academic Words	
photorealistic (a.)	逼真的，有真实感的
configuration (n.)	布局，构造；配置
algorithm (n.)	算法
generative (a.)	（语言学）生成的
adversarial (a.)	对抗的；对手的，敌手的
random (a.)	任意的，随机的，胡乱的
reliably (adv.)	可靠地；确实地
reconfigurable (a.)	可重构的
consistent (a.)	始终如一的，一贯的；持续的
layout (n.)	布局，设计
rectangular (n.)	长方形的，矩形的
bounding (n.)	边界，边界框
proportioned (a.)	相称的；成比例的
silhouette (n.)	(浅色背景衬托出的) 暗色轮廓；剪影
wintry (a.)	寒冷的，冬天的；冷淡的
resemble (v.)	像，与……相似
impressionistic (a.)	印象派的
distorted (a.)	扭曲的，变形的
benchmark (n.)	基准；（问题）标准检查程序
component (n.)	组成部分，成分，部件

supervised (a.)	有监督的
harness (v.)	控制并利用
convert (v.)	（使）转变，（使）转换
transparency (n.)	透明，透明性
scratch (v.)	划破；划出，刮出（痕迹）
envision (v.)	想象，预想
scenario (n.)	场景，可能发生的情况
autonomous (a.)	自主的，有自主权的

B. Words and Phrases

Exercises 1　Word Choice

Use the words in the box to finish the sentences.

configuration	algorithm	generative	adversarial	generate
rectangular	silhouette	synthesize	scratch	envision

1. We spent an hour or two in class learning how to _____ the codes, and in the end, everything gets easier.

2. Be careful not to _____ the furniture.

3. They _____ an equal society, free from poverty and disease.

4. A vitamin is a chemical compound that cannot be _____ by the human body.

5. The algorithm that eventually created an original image had two parts that worked against each other, called the Generator and the Discriminator; they dubbed this combative AI _____ Adversarial Network (GAN), Caselles-Dupré explained.

Exercise 2　Phrases

Match the words provided below with appropriate one in the box.

1. dark _____	box
2. learning _____	Silhouette
3. particular _____	Tone
4. rectangular _____	Algorithms
5. adversarial _____	Configuration

Exercise 3　Sentence Completion

Complete the sentences by filling in the blanks with phrases in the above exercise.

1. Many robots are equipped with high-tech sensors and complex _____ to avoid injuring humans as they work side by side.

2. An _____ is not in any way productive.

3. He put a _____ on the table.

4. Prices of this brand of computers range from $119 to $199, depending on the _____.

5. The _____ of the castle ruins stood out boldly against the fading light.

Exercise 4   Translation

Translate the sentences by using the words and phrases you have learned in the above two exercises.

1. 人工智能模型需要识别不同<u>排列组合</u>的物体，并将它们创建成一个图像。
_____

2. 这的确是很多行业分析师们所<u>预见</u>的情形。
_____

3. 与其他人工智能模型相比，LostGANs 能够<u>合成</u>像素更高的图像。
_____

4. 城堡遗迹的<u>黑色轮廓</u>在暮色映衬下清晰可见。
_____

5. 商家和消费者之间几乎存在一种<u>敌对</u>情绪。
_____

C. Collocation

Exercise 1   Modifiers

Find out the adjectives that modify the noun "image" in the article. The first letter has been provided.

p _____ image (para. 1)

c _____ image (para. 1)

c _____ image (para. 2)

r _____ image (para. 2)

c _____ image (para. 2)

r _____ image (para. 2)

Exercise 2   Blank Filling

Scan the text and complete the sentences containing the word "image".

1. AI has already shown off the capability to _____ images of cats, dogs, and people's faces that never existed before.

2. …how to train AI models to _____ images that could include many different objects…

3. …to generate more controlled images based on certain conditions rather than simply _____ image.

4. … we can _____ an image that can be consistent……

5. … previous AI models have generally failed to _____ images.

6. But LostGANs can _____ images at a resolution of up to 512 × 512 pixels…

7. …using fewer labeled conditions without having to sacrifice the quality of the _____ image.

8. LostGANs attracted the attention of local high school students with its still imperfect _____ images.

Exercise 3   Translation

Translate the sentences below from Chinese to English using "image" and its collocation in this section.

1. 你可以把这些图像存储在单独的计算机文件夹中。
   _____

2. 这项技术能够显示用于呈现数字图像的像素信息。
   _____

3. 视觉图像正在逐步取代文字。
   _____

4. 我画了一个网格，并把它粘贴起来创建图像。
   _____

5. 它还可以被用来从正在运行的安装中生成图像，包括所有数据和应用，用于备份。
   _____

D. Terminology

Exercise 1   Table Filling

Read the article and find the English technical terms according to the Chinese equivalents.

English Technical Terms	Chinese Equivalents
	深度学习算法
	生成对抗网络
	半监督学习
	视觉场景
	物体轮廓
	矩形边界框

Exercises 2   Blank Filling

Use the terms in the above exercise to complete the sentences below.

1. The challenge involves figuring out how to get AI models—in this case typically a class of _____ known as _____…

2. Such _____ can rely on just 50 percent of the usual training images to bring LostGANs up to its usual performance standards.

3. Such image generation based on _____ reconfigurable layouts could also potentially help generate different _____ that could help train autonomous vehicles.

4. This intermediate "layer-to-mask" step allows the model to further refine the general shape of such _____, which helps to make a more realistic and final "mask-to-image" result where all the visual details have been filled in.

5. The starting point for the LostGANs approach involves a simple reconfigurable layout that includes _____ showing where a tree, road, bus, sky, or person should be within the overall image.

Critical Reading and Writing

A. Brainstorming

Team up in a group with 3-5 students. Brainstorm and discuss "How can AI-generated images be used in industries such as advertising, entertainment, or healthcare?"

Industries	Applications of AI Image Generation
Advertising	
Entertainment	
Healthcare	

B. Critical Reading

The following are the answers ChatGPT provides for how AI-generated images can be used in advertising. Read and match the headings with each aspect.

1. Dynamic Content Generation;
2. Personalized Advertising;
3. Product Visualization.

⑤ ChatGPT

AI-generated images can be used in advertising in several ways:

   a. _____: AI can generate realistic images of products, allowing advertisers to display their products in various settings without the need for costly photoshoots. This enables brands to present their products in nice places that their customers like.

   b. _____: AI algorithms can analyze consumer data to create personalized images tailored to individual preferences. This allows advertisers to deliver more relevant and engaging content that people like and want to buy.

   c. _____: AI can generate dynamic images that change based on contextual factors such as location, time of day, or weather conditions. This allows advertisers to make ads that catch your eye and feel like they're made just for you.

Exercise 2: Work in a group. What are the applications of AI image generation in the other two industries? You may choose either entertainment or healthcare to develop your ideas. Follow the patterns below.

Entertainment
1. AI can .... This allows/enables ...
2. AI can .... This allows/enables ...
3. AI can .... This allows/enables ...

Healthcare
1. AI can .... This allows/enables ...
2. AI can .... This allows/enables ...
3. AI can .... This allows/enables ...

C. Essay Writing

Having engaged in discussions and vocabulary preparation, you've likely generated numerous insightful ideas. Now, it's time to reflect on these ideas and the knowledge you've acquired by crafting an essay. Use the following instructions to guide your writing process:

Topic
- Benefits and challenges of AI-generated images in the entertainment industry.
- Benefits and challenges of AI-generated images in the healthcare industry.

Background Information

Artificial Intelligence(AI) has transformed various industries, enabling machines to perform tasks traditionally requiring human intelligence. One significant application is the generation of images using advanced machine learning techniques like Generative Adversarial Networks(GANs). These AI-generated images are gaining popularity in entertainment and healthcare for their ability to create realistic visuals autonomously. However, their use also raises ethical and technical challenges, such as data privacy concerns and algorithm bias. Despite these challenges, AI-generated images hold vast potential to revolutionize industries and improve visual content creation and analysis.

Instructions

In an essay of approximately 300-450 words, <u>analyze the applications of AI image</u>

generation technology in either entertainment industry or health industry. Provide specific evidence and arguments to support your opinion.

Your essay should include the following components:

Introduction (approximately 50-75 words): Briefly introduce the topic and provide context for your analysis.

- Provide an overview of the topic and introduce the importance of AI-generated images in the industry you have chosen to discuss.

Body Paragraphs (approximately 250-350 words): Present your argument in detail, supported by specific evidence and examples in 3 separate paragraphs.

- Discuss the potential benefits of using AI-generated images in the industry, such as improved efficiency, cost-effectiveness, or enhanced user experiences.
- Discuss potential limitations or risks and propose strategies for addressing them.
- Future prospects: Offer insights into the future of AI-generated images in the industry and speculate on potential advancements or innovations that may emerge.

Conclusion (approximately 50-75 words): Summarize your main points and restate your thesis in light of the evidence presented.

- Summarize the key points of your essay and provide a concluding statement that reinforces the significance of AI-generated images in shaping the future of the chosen industry.

* Ensure that your essay is well-structured, logically organized, and supported by evidence from reputable sources. Use clear and concise language, and proofread your work carefully for grammar, punctuation, and spelling errors.

Discussion and Presentation

A. Group Discussion

Exercise 1: Think and discuss the questions below.

1. How is AI being utilized in China's healthcare system, and what are the main areas of focus for AI applications in healthcare?

2. What role does computer vision play in AI-driven healthcare solutions, according to Hsiao-wuen Hon, the managing director of Microsoft Research Asia?

3. How are training and education playing a crucial role in the integration of AI with healthcare in China, and what are the key initiatives being undertaken to address this?

Exercise 2: Read the article below. Search for any information related to the above questions. Do you think the answers provided in the article match yours? Share your ideas with the class.

B. Extended Reading-AI in China

Read the article, highlight any ideas or language that will help you in your presentation in the next session.

AI in Healthcare: China's Quest for Better Healthcare with AI

In China, healthcare has become one of the largest applications for Artificial Intelligence (AI). Companies from start-ups to tech giants are seizing the opportunity to apply AI solutions.

A rapidly aging population is putting a lot of stress on China's healthcare system. The Chinese government has been looking to technology to solve the problem. One of the areas it sees real potential and room for growth is AI, with most of China's AI healthcare companies currently focused on medical imaging systems.

Hsiao-wuen Hon, the managing director of Microsoft Research Asia, said, "As we know, the computer vision is a big part of AI. And in healthcare, looking of the image whether it is the X-ray or MRI, we need to get a special doctor spending lots of time to look at the image. Computer vision can provide a lot of help. It's not really trying to replace the doctor, but really making doctor productive. You have already a lot of applications going on in the medical imaging."

As AI becomes increasingly sophisticated and is integrated more with healthcare, training becomes increasingly important. Special multi-form platforms are being built to meet this end, for both full-time and reserve personnel. Experts say this will provide more adequate support for scientific and industrial development.

"Workers also need to upgrade their knowledge in AI. We also will use this platform to provide the training certification to make sure the existing workers are well educated and can process necessary knowledge in AI," stressed Hon.

Experts say while advancements in the industry could alleviate China's personnel shortage, its diagnostic accuracy still needs improvement.

C. Presentation

After you have read the article, please choose one of the following topics to develop your ideas. Make a presentation with PowerPoint to the class.

Topic 1   AI Revolution in China's Healthcare: Opportunities and Challenges
- Provide an overview of the current landscape of AI applications in China's healthcare.
- Discuss the opportunities AI presents for addressing challenges such as an aging population and personnel shortages in the healthcare system.
- Explore the challenges and ethical considerations associated with integrating AI into healthcare practices.
- Provide examples of AI technologies currently being utilized in China's healthcare sector and discuss potential future developments.

Topic 2   Harnessing Computer Vision in Healthcare: The Role of AI in Medical Imaging Systems
- Focus on the specific application of AI, particularly computer vision, in medical imaging systems.
- Discuss the benefits of AI in improving diagnostic accuracy, increasing efficiency, and reducing the workload on healthcare professionals.
- Address concerns related to data privacy, algorithm bias, and the need for specialized training for healthcare personnel. The presentation should include real-world examples and case studies to illustrate the impact of AI in medical imaging.

Warm Up

Exercise 1: Video Watching

Note:
- Craft a visually appealing PowerPoint with appropriate colors and images.
- Keep each slide concise, using fewer than 10 words.

- Use Pictures, illustrations or forms to make your point.
- Emphasize positive concepts and messages throughout your presentation.
- Aim to deliver your presentation smoothly, without relying on notes, within a timeframe of 4-5 minutes.

Video

A. Before You Watch

Read out the words below. Choose a word or a phrase in the box to form an appropriate expression.

water   image   map   oil painting   images   culture   high

- compelling _____
- _____ reflection
- segmentation _____
- _____ texture
- synthesize _____
- _____ diversity
- _____ fidelity

B. While You Watch

Exercise 1: Discussion

Considering the capabilities of NVIDIA'S GauGAN, how do you think this technology could revolutionize the fields of digital art, virtual world creation, and architectural design?

Exercise 2: Dictation

Fill in the blanks with words and expressions you have heard from the video.

Wouldn't it be great if everybody could be an artist if we could take our ideas and turn them into (1)_____? This technology allows us to create (2)_____ so that if you want to create a new picture, you can just draw the shapes of the objects that you want, and the neo-networks can then fill in all the details.

If we add a water feature, the network is able to (3)_____, not because we told it that, but because it's learned it. Or if we change the ground to be covered with snow, then it knows that the sky also needs to be a different color. I really think this technology is going to be great for (4)_____, designers, people making (5)_____ to train robots and self- driving cars.

The input of this model is something we called the (6) _____. It's like a coloring book picture that describes here's where a tree is, here's where the sky is, here's where the ground is. And it doesn't have any details and then the neo-network is able to (7)_____ and the shadows and the colors based on things that it has learned from a large database of real-world images.

The real advance here is that we were able to (8)_____ images with a lot more (9)_____ and more (10)_____ than we were able to in the past. I really think this technology is gonna be great for the dreamers of the world.

Summary and Reflection

Now that you have completed the chapter of AI image generation technology, it's time to reflect on your learning and ensure you have met the goals set for the chapter. Follow these steps to complete the checklist:

- Carefully read through the checklist provided, which outlines the key learning objectives and goals of the AI image generation technology chapter. For each item on the checklist, evaluate your own understanding and progress by checking the corresponding box.
- If you feel confident in your understanding and achievement of the goal, check the box; If you believe there are areas where you need further improvement or clarification, leave the box unchecked.

1. **Understanding of AI Image Generation Technology:**
   ☐ Have I deepened my understanding of AI image generation technology, including the underlying machine learning techniques and algorithms used?
   ☐ Can I explain the process of AI image generation clearly to others, including concepts such as segmentation maps and neural networks?
2. **Critical Thinking Skills:**
   ☐ How effectively did I apply critical thinking skills to tasks such as analyzing the feasibility and implications of AI-generated images in different scenarios?

☐ Did I identify potential limitations or ethical considerations associated with the use of AI image generation technology?

3. **Reading Comprehension, Vocabulary, and Language Proficiency:**

   ☐ Have I improved my reading comprehension, vocabulary, and language proficiency through exercises focused on technical texts and terminology related to AI image generation?

   ☐ Can I accurately define and use technical terms associated with AI image generation in context?

4. **Collocating Words and Phrases:**

   ☐ Can I identify and analyze the collocation of words and phrases within the context of AI image generation to enhance my understanding of technical terminology?

   ☐ Have I practiced using these collocations in sentences to reinforce my understanding?

5. **Understanding of Impacts:**

   ☐ How did I explore the potential impacts of AI image generation technology on industries such as entertainment, healthcare, and advertising?

   ☐ Can I articulate the broader societal implications of widespread adoption of AI-generated images?

6. **Presentation Skills:**

   ☐ How did I develop my presentation skills by creating and delivering presentations on topics related to AI image generation, using tools like PowerPoint?

   ☐ Did I effectively organize information, use visual aids, and engage the audience to convey complex concepts?

7. **Reflection and Critical Analysis:**

   ☐ How did I reflect on and critically analyze the knowledge acquired throughout the chapter, particularly in relation to the ethical considerations and future applications of AI image generation?

   ☐ Did I consider different perspectives and evaluate the significance of AI image generation advancements in various fields?

Chapter 4
Robot

> **Objectives**
> In this chapter, you should be able to:
> - Analyze and discuss advances and potential impacts of robotics.
> - Develop critical thinking, reading, and writing skills by analyzing the development and challenges of robot technology.
> - Demonstrate the ability to summarize and categorize information from complex texts related to robot technology using techniques such as multiple choice, mind mapping, and matching exercises.
> - Expand vocabulary and language proficiency in an academic context through exercises like glossaries, word choice, and translation tasks related to robot technology and its terminology.
> - Engage in collaborative discussions and presentations to share insights and perspectives on robot technology, its impacts and innovation, both before and after reading relevant material.

 Before You Read

A. Discussion

Look at the pictures below and discuss with a partner.
1. What are the functions of the robots in the pictures?
2. Which kind of robot will be most popular in the future?

B. Skimming and Scanning

Browse the text and answer the questions below.

1. What is the PR2 NOT capable of?

A. Bringing a bowl

B. Pouring milk into a cup

C. Tossing the milk box into the trash

D. Putting the cereal box back into its storage location

2. How long did the PR2 take to complete all the tasks set in the competition?

A. 20 minutes

B. 50 minutes

C. 70 minutes

D. 90 minutes

3. How many times did PR2 succeed in the table setting tasks?

A. 1

B. 3
C. 5
D. 7

 Text

It's (Still) Really Hard for Robots to Autonomously Do Household Chores

Something as Simple as Breakfast Takes This PR2 90 Minutes to Set Up
and Then Clean, and It's Not Always Successful

1. Every time we think that we're getting a little bit closer to a household robot, new research comes out showing just how far we have to go. Certainly, we've seen lots of progress in specific areas like grasping and semantic understanding and whatnot, but putting it all together into a hardware platform that can actually get stuff done autonomously still seems quite a way off.

2. In a paper presented at ICRA 2021 this month, researchers from the University of Bremen conducted a "Robot Household Marathon Experiment," where a PR2 was tasked with first setting a table for a simple breakfast and then cleaning up afterwards in order to "investigate and evaluate the scalability and the robustness aspects of mobile manipulation." While this sort of thing kinda seems like something robots should have figured out, it may not surprise you to learn that it's actually still a significant challenge.

3. PR2's job here is to prepare breakfast by bringing a bowl, a spoon, a cup, a milk box, and a box of cereal to a dining table. After breakfast, the PR2 then has to place washable objects into the dishwasher, put the cereal box back into its storage location, toss the milk box into the trash. The objects vary in shape and appearance, and the robot is only given symbolic descriptions of object locations (in the fridge, on the counter). It's a not only very realistic but also very challenging scenario, which probably explains why it takes the poor PR2 90 minutes to complete it.

4. First off, kudos to that PR2 for still doing solid robotics research, right? And this research is definitely solid—the fact that all of this stuff works as well as it does, perception,

motion planning, grasping, high level strategizing, is incredibly impressive. Remember, this is 90 minutes of full autonomy doing tasks that are relatively complex in an environment that's not only semi-structured and somewhat, but not overly, robot-optimized. In fact, over five trials, the robot succeeded in the table setting task five times. It wasn't flawless, and the PR2 did have particular trouble with grasping tricky objects like the spoon, but the framework that the researchers developed was able to successfully recover from every single failure by tweaking parameters and retrying the failed action. Arguably, failing a lot but being able to recover a lot is even more useful than not failing at all, if you think long term.

5. The cleanup task was more difficult for the PR2, and it suffered unrecoverable failures during two of the five trials. The paper describes what happened:

6. Cleaning the table was more challenging than table setting, due to the use of the dishwasher and the difficulty of sideways grasping objects located far away from the edge of the table. In two out of the five runs we encountered an unrecoverable failure. In one of the runs, due to the instability of the grasping trajectory and the robot not tracking it perfectly, the fingers of the robot ended up pushing the milk away during grasping, which resulted in a very unstable grasp. As a result, the box fell to the ground in the carrying phase. Although during the table setting the robot was able to pick up a toppled over cup and successfully bring it to the table, picking up the milk box from the ground was impossible for the PR2. The other unrecoverable failure was the dishwasher grid getting stuck in PR2's finger. Another major failure happened when placing the cereal box into its vertical drawer, which was difficult because the robot had to reach very high and approach its joint limits. When the gripper opened, the box fell on a side in the shelf, which resulted in it being crushed when the drawer was closed.

7. While we're focusing a little bit on the failures here, that's really just to illustrate the exceptionally challenging edge cases that the robot encountered. Again, I want to emphasize that while the PR2 was not successful all the time, its performance over 90 minutes of fully autonomous operation is still very impressive. And I really appreciate that the researchers committed to an experiment like this, putting their robot into a practical environment doing practical tasks under full autonomy over a long period of time. We often see lots of incremental research headed in this general direction, but it'll take a lot

more work like we're seeing here for robots to get real-world useful enough to reliably handle those critical breakfast tasks.

 Reading Comprehension

A. Multiple Choice

Choose the best answer for each question.

1. What is the main idea of the passage?

A. The PR2 has fully achieved the goal of household tasks.

B. Household robots still have a long way to go before they can be practical.

C. The PR2 is the most advanced household robot in the world.

D. Researchers have solved all the challenges in household robotics.

2. What caused the milk box to fall to the ground?

A. The robot's fingers were too weak.

B. The grasping trajectory was unstable.

C. The robot was not able to track the trajectory.

D. The robot's grippers were too small.

3. Which of the following is NOT a failure that the PR2 encountered?

A. Falling over while carrying the milk box.

B. Failing to grasp the spoon.

C. The dishwasher grid getting stuck.

D. The cereal box being crushed in the drawer.

4. What does the author think of the PR2's performance?

A. It is disappointing.

B. It is mediocre.

C. It is impressive.

D. It is unpredictable.

5. What can we infer from the passage?

A. Household robots will soon be able to replace humans.

B. The PR2 is the most suitable for household tasks.

C. More research is needed for household robots to be useful.

D. The PR2 is the only one that can do household tasks.

B. Mind Map

How many main parts do you think the article is composed of? Group the paragraphs and fill in the blanks with the information you read from the article.

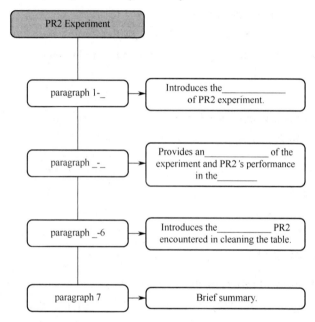

C. Matching

Read the text and decide which paragraph mentions the following information? Write the number of the paragraph before each sentence.

_____ 1. The PR2 was tasked with preparing breakfast and cleanup, which took it 90 minutes to complete due to the complexity of grasping objects with varying shapes and descriptions.

_____ 2. Despite encountering challenges, the PR2 demonstrated impressive capabilities in perception, motion planning, grasping, and high-level strategizing, successfully recovering from failures in most cases.

_____ 3. Every time advancements in household robots are anticipated, new research highlights the significant gap between current capabilities and the ideal.

_____ 4. The paper describes the specific challenges and failures encountered by the

PR2 during the cleanup task, highlighting difficulties with grasping and manipulating objects in a realistic environment.

_____ 5. Researchers at ICRA 2021 conducted a marathon experiment using a PR2 to test its ability to set a breakfast table and clean up afterwards, revealing challenges in mobile manipulation.

_____ 6. While focusing on the failures, it is important to note the PR2's overall impressive performance in 90 minutes of fully autonomous operation, demonstrating the progress made in practical robotics research.

_____ 7. The cleanup task was more difficult for the PR2, resulting in two unrecoverable failures due to challenges with grasping and placing objects accurately.

D. Cloze

The information below is a summary from the text. Complete the summary by filling in the blanks with the words and phrases provided.

A. shape and appearance	F. object locations
B. semantic understanding	G. fully autonomous operation
C. a practical environment	H. grasping tricky objects
D. instability	I. mobile manipulation
E. the dishwasher grid	J. semi-structured

Nowadays, household robots have made great progress in some specific fields, such as grasping and (1)_____. But they still have a lot to be improved. Researchers from the University of Bremen published a paper on ICRA 2021 about "Robot Household Marathon Experiment". This experiment aims to analyze the scalability and the robustness aspects of (2)_____.

The experiment can be divided into two parts, which are first setting a table for a simple breakfast and then cleaning up afterwards. All the objects PR2 should work on are different in (3)_____ but PR2 is only provided with the description of (4)_____. PR2 spent 90 minutes since the task is really challenging. The ability of the PR2 to autonomously perform a 90-minute task in a (5)_____ environment is remarkable. PR2's performance in table-setting is not perfect when (6)_____ but every single failure was recovered by tweaking parameters. As a result, during the table-setting task, PR2 is successful in all the five tests.

For the PR2, the cleanup was more difficult than the table-setting, with two out of five tests showing unrecoverable problems. In one of the failures, the main reason of grasping failure is the (7)_____ of the grasping trajectory. Another problem that can't be fixed is that (8)_____ got stuck on the PR2's finger.

It's worth emphasizing that even though the PR2 isn't always a success, its 90 minutes of (9)_____ is still incredible. Researchers also deserve our admiration for putting their robots in (10)_____ and performing practical tasks with complete autonomy for a long period of time.

 # Language Building

A. Glossary

Proper Nouns
PR2（Personal Robot 2） 个人机器人 2，是威楼加拉吉生产的机器人。它的前身是斯坦福研究生埃里克·伯格和基南·威罗拜克开发的 PR1 机器人。
ICRA ICRA 全称 IEEE International Conference on Robotics and Automation，即 IEEE 国际机器人技术与自动化会议，其是机器人学领域一年一度的学术会议，是这一领域的早期会议之一。
University of Bremen 不来梅大学，成立于 1971 年，位于德国不来梅市，是德国著名的较"年轻"的大学之一。

Academic Words	
semantic (a.)	语义的
whatnot (n.)	诸如此类
scalability (n.)	可扩展性；可伸缩性；可测量性
robustness (n.)	鲁棒性，稳健性；健壮性
manipulation (n.)	操纵；控制
kinda (adv.)	有点，有几分
toss (v.)	扔，抛，掷
scenario (n.)	场景
kudos (n.)	荣誉

perception (n.)	看法，认识；感觉，感知
incredibly (adv.)	难以置信地；非常地
autonomy (n.)	自主性；自主
optimize (v.)	优化
flawless (a.)	完美的，无瑕的
tricky (a.)	难对付的，棘手的
tweak (v.)	稍做调整（机器、系统等）
parameter (n.)	参数，变量
arguably (adv.)	按理说
trial (n.)	试验
instability (n.)	不稳定性
trajectory (n.)	轨道，轨迹
phase (n.)	阶段，时期
topple (v.)	（使）不稳而倒下
vertical (a.)	直立的
committed (a.)	尽心尽力的
incremental (a.)	递增的，增量的

B. Words and Phrases

Exercises 1   Word Choice

Use the words in the box to finish the sentences.

autonomously	scalability	robustness	manipulation	scenario
perception	trial	tweak	parameter	flawless

1. Unfortunately, cloud computing's track record to date has hardly been _____.

2. More than the physical reality of photons or sound waves, _____ is a product of the brain

3. In the worst-case _____ more than ten thousand people might be affected.

4. If it's a straight line, that's quite good: "linear _____".

5. That's because there are a number of powerful voice _____ and automation technologies that are about to become widely available for anyone to use.

Exercise 2   Phrases

Match the words provided below with appropriate one in the box.

1. autonomous _____
2. grasping _____
3. _____ research
4. semi _____
5. motion _____

| planning |
| incremental |
| operation |
| trajectory |
| structured |

Exercise 3　Sentence Completion
Complete the sentences by filling in the blanks with phrases in the above exercise.

1. And what do you do if the amount of structured, _____ and unstructured data are roughly equal?

2. Mobile robot _____ is one of the key technologies in the robot navigation.

3. Space robot control language is the software foundation for realizing three control modes of space robot: remote control, _____ and cooperative operation.

4. The _____ of the robot arm is an important part of the preset program.

5. In _____, it can reduce waste and promote the development of science and technology.

Exercise 4　Translation
Translate the sentences by using the words and phrases you have learned in the above two exercises.

1. 他期望在工程师对该系统的性能稍加改进后，它会运行得更好。
_____

2. 近年来，机器人运动规划问题渐渐成为机器人学研究中的热点问题。
_____

3. 上面的概念同样适用于数据操作。
_____

4. 为了使这种差异的效果最大化，我们对操作参数进行了调整。
_____

5. 如果小鸟毁坏了一部发动机，则最坏的情形是飞机坠毁。
_____

C. Collocation

Exercise 1　Modifiers
Find out the adjectives and verbs using with the noun "failure" in the article. The first letter has been provided.

s_____ failure (para.4)
s_____ failure (para.5)
e_____ failure (para.6)
u_____ failure (para.6)
m_____ failure (para.6)

Exercise 2 Blank Filling

Scan the text and complete the sentences containing the word "failure".

1. In two out of the five runs we _____ an unrecoverable failure.

2. ... but the framework that the researchers developed was able to successfully recover from every _____ failure by tweaking parameters and retrying the failed action.

3. The other _____ failure was the dishwasher grid getting stuck in PR2's finger.

4. The cleanup task was more difficult for the PR2, and it _____ unrecoverable failures during two of the five trials.

5. Another _____ failure happened when placing the cereal box into its vertical drawer, ...

Exercise 3 Translation

Translate the sentences below from Chinese to English using "failure" and its collocation in this section.

1. 这对大多数病人有用，但是很多病人得了<u>心力衰竭</u>，疾病已向脏器扩散。

_____

2. 因此当他们遭遇<u>失败</u>的时候，他们就崩溃了。

_____

3. <u>设备故障</u>可发生在工件加工时和加工前。

_____

4. 有<u>重大故障</u>时，可与本厂联系。

_____

5. 防止<u>市场失灵</u>的政策已经普遍实施。

_____

D. Terminology

Exercise 1 Table Filling

Read the article and find the English technical terms according to the Chinese equivalents.

English Technical Terms	Chinese Equivalents
	鲁棒性
	运动规划
	优化
	抓取轨迹
	自动化操作
	增量研究

Exercises 2

Use the terms in the above exercise to complete the sentences below.

1. .... where a PR2 was tasked with first setting a table for a simple breakfast and then cleaning up afterwards in order to "investigate and evaluate the scalability and the _____ aspects of mobile manipulation."

2. Remember, this is 90 minutes of full autonomy doing tasks that are relatively complex in an environment that's not only semi-structured and somewhat, but not overly, robot-_____.

3. We often see lots of _____ headed in this general direction, but it'll take a lot more work like we're seeing here for robots to get real-world useful enough to reliably handle those critical breakfast tasks.

4. In one of the runs, due to the instability of the _____ and the robot not tracking it perfectly, the fingers of the robot ended up pushing the milk away during grasping, which resulted in a very unstable grasp.

5. Again, I want to emphasize that while the PR2 was not successful all the time, its performance over 90 minutes of fully _____ is still very impressive.

6. And this research is definitely solid—the fact that all of this stuff works as well as it does, perception, _____ , grasping, high level strategizing, is incredibly impressive.

 # Critical Reading and Writing

A. Brainstorming

Work in groups. Fill in the table according to the instruction.

Although robots will bring all kinds of convenience to our lives, it will still bring some negative effects. Please list three negative effects of robots and indicate the reasons.

Negative Effects	Reasons
1	1
2	2
3	3

B. Critical Reading

Exercise 1: The following are the answers to the question provided by ERNIE Bot. Read the introduction, and then match the headings and the corresponding content in the body part.

Heading A: Threats to Privacy and Security

Heading B: Exacerbation of Economic Inequality

Heading C: Dependency and Increased Laziness

Heading D: Alienation of Social Relationships

Heading E: Moral and Ethical Issue

Heading F: Strike on Job Market

With the rapid development of technology, robotic technology has penetrated into every aspect of our lives, ranging from industrial production to household services, from medical fields to military applications. However, the proliferation of robots has also brought a series of adverse impacts that deserve our deep consideration.

a. _____:

The widespread application of robots has led to the replacement of many traditional jobs by automation, causing significant disruption to the job market. This is particularly true in the manufacturing and logistics sectors, where robots are already capable of handling a large number of repetitive and labor-intensive tasks, putting many workers at risk of unemployment. Moreover, as technology continues to advance, robots will be able to perform more complex tasks, further intensifying competition in the job market.

b. _____:

As robots become more prevalent, more of them are being used to collect and process personal information. This can increase the risk of personal privacy breaches, especially in the case of household service robots and intelligent assistants. Additionally, malicious attackers may exploit robotic technology to conduct cyber attacks or physical attacks, posing threats to personal and societal security.

c. _____:

As robots take on more social and entertainment functions, people may gradually become accustomed to interacting with robots and neglect establishing connections with real-world individuals. This can lead to a decline in social skills and even trigger

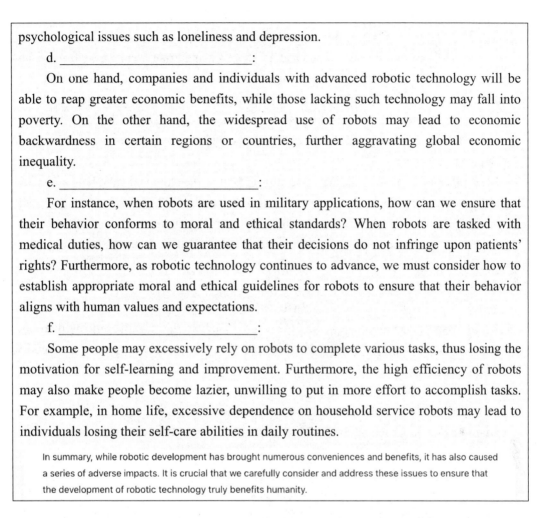

psychological issues such as loneliness and depression.

  d. _____:

  On one hand, companies and individuals with advanced robotic technology will be able to reap greater economic benefits, while those lacking such technology may fall into poverty. On the other hand, the widespread use of robots may lead to economic backwardness in certain regions or countries, further aggravating global economic inequality.

  e. _____:

  For instance, when robots are used in military applications, how can we ensure that their behavior conforms to moral and ethical standards? When robots are tasked with medical duties, how can we guarantee that their decisions do not infringe upon patients' rights? Furthermore, as robotic technology continues to advance, we must consider how to establish appropriate moral and ethical guidelines for robots to ensure that their behavior aligns with human values and expectations.

  f. _____:

  Some people may excessively rely on robots to complete various tasks, thus losing the motivation for self-learning and improvement. Furthermore, the high efficiency of robots may also make people become lazier, unwilling to put in more effort to accomplish tasks. For example, in home life, excessive dependence on household service robots may lead to individuals losing their self-care abilities in daily routines.

  In summary, while robotic development has brought numerous conveniences and benefits, it has also caused a series of adverse impacts. It is crucial that we carefully consider and address these issues to ensure that the development of robotic technology truly benefits humanity.

Exercise 2: Do you agree with ERNIE Bot on the above ideas? Can you think of any more aspects? Can you think of solutions to these problems? Share your ideas with the class.

C. Essay Writing

Having engaged in discussions and vocabulary preparation, you've likely generated numerous insightful ideas. Now, it's time to reflect on these ideas and the knowledge you've acquired by crafting an essay. Use the following instructions to guide your writing process:

● **Topic**

The Negative Effect of the Development of Robotics on … (a specific aspect)

Background Information

As technology continues to advance at an unprecedented rate, robots have become an integral part of our daily lives. They change the way we work, learn, communicate, and even connect with each other. However, while the integration of robots into various sectors has brought many benefits, such as increased efficiency, accuracy, and reduced costs, it has also raised concerns about the potential negative effects that accompany such rapid development.

The topic of exploring the negative impact of the development of robotics and its impact on society is particularly important in today's world, as robots are increasingly used in everything from manufacturing and healthcare to transportation and entertainment. As we increasingly rely on robots to perform tasks, it's critical to understand the potential drawbacks and challenges that may arise.

Instructions

In an essay of approximately 300-450 words, analyze the negative effects of robot development on a particular aspect. Provide specific evidence and arguments to support your opinion.

Your essay should include the following components:

Introduction (approximately 50-75 words): Briefly introduce the topic and provide context for your analysis.

- State your thesis or main argument regarding what negative effects are caused by the robot development on a particular aspect.

Body Paragraphs (approximately 250-350 words): Present your argument in detail, supported by specific evidence and examples in 3 separate paragraphs.

- Consider the differences between robots and humans.
- Explore the reasons for the negative impact of robots in specific aspects.
- Present ideas from different angles.

Conclusion (approximately 50-75 words): Summarize your main points and restate your thesis in light of the evidence presented.

- Highlight the negative effects of the development of robotics and call for attention and solutions to these problems.
- Put forward the prospects and suggestions for the development of robot technology in the future.

> * Ensure that your essay is well-structured, logically organized, and supported by evidence from reputable sources. Use clear and concise language, and proofread your work carefully for grammar, punctuation, and spelling errors.

Discussion and Presentation

A. Group Discussion

Exercise 1: Think and discuss the questions below.

1. What significant achievements have China made in independent innovation in robot technology, and how have these achievements elevated China's position in the global robotics industry?

2. What are the major challenges and opportunities faced by China's robotics industry during its development, and how can it overcome the challenges and seize the opportunities?

3. How will the integration of artificial intelligence and robot technology drive the development of robotics, and what impact will this integration have on China's robotics industry?

4. What role have international cooperation and exchanges played in promoting the development of China's robot technology, and what are some successful cooperation cases? How has the rise of China's robotics industry impacted the global competitive landscape?

Exercise 2: Read the article below. Search for any information related to the above questions. Do you think the answers provided in the article match yours? Share your ideas with the class.

B. Extended Reading—Robots in China

Read the article, highlight any ideas or language that will help you in your presentation in the next session.

Innovation Brings Chinese Robots to the World

China now has the largest robot application market in the world, according to Xin Guobin, vice minister of industry and information technology, speaking at a forum during the World Robot Conference 2022 (WRC 2022).

Rising to Robot R&D Challenges

A key factor of China's progress in the robotics industry is the joint efforts of universities, research institutes and enterprises in tackling technological challenges in robot R&D.

Breakthroughs of technologies regarding key parts of robots, such as reducer, controller and servo systems, have been achieved, especially in enterprises, among which is SIASUN, a robotic enterprise belonging to the Chinese Academy of Sciences.

As a leading enterprise in Automatic Guided Vehicle (AGV) SIASUN was among the first enterprises to apply AGV in real life situations like automobile assembling. The synchronous tracking technology developed by SIASUN can dynamically follow the passive reflector of the hoist for car body lifting, thus making dynamic assembling possible by checking the deviation of relative positioning.

Applying AGV in heavy truck assembling is a more recent challenge tackled by SIASUN. Different from the demand for AGV in car assembling, heavy truck assembling requires stronger bearing capacity, lifting ability and safety factors. To meet the needs, researchers in SIASUN achieved technological innovation in lifting structure, control unit and sensing unit through independent R&D.

There are also many other technological achievements in the robotics industry. Industrial robots have become more flexible after compliant force control features were added to them, which helps to realize their application in scenarios that require higher precision and stronger sensitivity.

Robots in Daily Life

Robots have deeply integrated into daily Chinese life. Industrial robots have been applied in 60 industry categories and 168 industries. According to the National Bureau of Statistics, the production volume of industrial robots in China in 2021 reached 366,000, soaring by 68 percent compared with that of the previous year.

Industrial robots are key to intelligent manufacturing systems, which play an important role in the digital transformation of the manufacturing industry. The application of industrial robots has also expanded from simple tasks like carrying materials, to precision machining like polishing.

Delivery robots are now a more common sight, along with housework robots. Meanwhile surgical robots are of great help to doctors in terms of completing highly difficult surgeries with precision, efficiency and safety, as they advance the limits of human hands, eyes and brain.

Also, special robots are widely adopted in emergency rescue, hazardous operations, and scientific expeditions in extreme conditions. Aerial work is quite dangerous as workers have to face both the risk of falling and extreme weather like strong wind, high temperature and severe cold. A special robot displayed at the expo during WRC 2022 can release workers from such situations and has a working efficiency six to eight times that of a human.

Chinese Robots Going Global

In addition to serving the domestic market, robots developed and produced by China can be found across the world.

As early as 2007, SIASUN won the bid in an AGV project from the U.S. company General Motors, which was the first time that China's AGV products had been exported to developed countries. Now, SIASUN's products have been sold to more than 40 countries and regions around the globe.

Chengdu CRP Robot Technology Co., Ltd., listed by the Ministry of Industry and Information Technology as a "little giant" among all the Specialized, Refined, Differential and Innovative(SRDI) enterprises in China in 2021, witnessed an increase of export orders by nearly 40 percent year-on-year for the first half of 2022.

The export value of industrial robots has experienced obvious growth for the past few years. According to China Customs, the export value of industrial robots reached about 3.67 billion RMB in 2021, surging by 40 percent compared with exports in 2020.

December 2021, 15 departments in China jointly issued a guideline for the robotics industry during the 14th Five-Year Plan period(2021—2025). The guideline says that the country aims to build 3-5 robotics industry clusters with international influence by 2025.

This augurs well for this industry given the projected ongoing growth of the global robot market.

C. Presentation

After you have read the article, please choose one of the following topics to develop your ideas. Make a presentation with PowerPoint to the class.

Topic 1 Technological Innovation and Development Trend of China's Robotics Industry
- Introduce the current status of technological innovation in China's robotics industry, including achievements and existing problems.
- Analyze the future development trends of China's robotics industry.
- Discuss the key technological breakthroughs made by China's robotics industry.
- Analyze the policy support provided by the Chinese government for technological innovation in the robotics industry.

Topic 2 The Competitiveness of Chinese Robotics Industry in the International Market
- Introduce the current competitive situation of China's robotics industry in the international market.
- Analyze the advantages and disadvantages of China's robotics industry in the international market.
- Discuss the competitive strategies of China's robotics industry in the international market.
- Analyze the opportunities and challenges of China's robotics industry in international cooperation.

Note:
- Craft a visually appealing PowerPoint with appropriate colors and images.
- Keep each slide concise, using fewer than 10 words.
- Use pictures, illustrations or forms to make your point.
- Emphasize positive concepts and messages throughout your presentation.
- Aim to deliver your presentation smoothly, without relying on notes, within a timeframe of 4-5 minutes.

Video

A. Before You Watch

Read out the words below. Choose a word in the box to form an appropriate expression.

self-driving   reinforcement   artificial   autonomous   science   mechanical

- _____ intelligence
- computer _____
- _____ tools
- _____ cars
- _____ robots
- _____ learning

B. While You Watch

Exercise 1: Discussion
What is the relationship between artificial intelligence and robotics?

Exercise 2: Dictation
Fill in the blanks with words and expressions you have heard from the video.

So basically, artificial intelligence is an area of (1) _____ that helps us to develop computer programs that can learn by themselves. So, you either feed them data and they learn from this data, or you use (2) _____ to help (3) _____. Robotics, on the other hand, is a field of engineering, that is, they basically focus on building and operating robots, so (4) _____ that can do things autonomously. So, we've seen this in manufacturing for a long time, where we have tools that can, and robots that can build cars, but we also now have very advanced robots that can do almost anything. And this is where the overlap between those two fields comes in, that we now can combine artificial intelligence and robotics, where basically the robot is the body, and the artificial intelligence is the brain.

So, if you think about this, in the past, we've had robots for a long-time building things like cars, but they've been dumb robots. They've picked something up, program to screw something into there, put a wheel onto a car, spray paint a car, but it couldn't really (5) _____. Nowadays, we can give those robots sensors, and we can give those robots things like cameras that act as their eyes, and we add artificial intelligence to this equation as a brain, and suddenly you have an (6) _____. And this is where all the advances have come from in robotics recently. If you think about a drone, e.g. as a robot, the brain, the artificial intelligence, makes this drone fly autonomously. We now have self-driving cars, again, combining robotics and the brain.

A really good example is a (7) _____. This is something I've looked at recently. And basically, in the past, it would have been impossible for a robot to pick raspberries. You think about this delicate food, and every bush looks different. The location of the raspberry is different, how to gently pick it. And people basically felt that you needed humans with their (8) _____ to do this. However, by now combining artificial intelligence and robotics, you can now achieve that. You now have a raspberry picking robot. Let users use machine vision or a camera to detect where the raspberries are, and then use an arm, a robotic arm with lots of sensors in it, to then pick the raspberry perfectly, not to push it too hard, and then place it somewhere in a container to have it harvested. And this, for me, is a really good illustration of how far autonomous robots have now come.

And basically, we now also have this ability of what we call (9) _____. So instead of feeding a machine lots of data to recognize a raspberry, you let the machine learn things by experience. Therefore, we now can use reinforcement learning. This is almost learning by experience. There fore, I've recently seen a robot that is basically monitoring the environment, that is learning by itself to start to walk. And this, each failure reinforces a learning point. Each successive was a successful step, and it the robot, e.g., didn't fall over. It then registers this. Thus, you now get this (10) _____ learning cycle where robots can almost learn like infants to do things like speaking and walking. And this, for me, is the latest evolution of these, of combining artificial intelligence and robotics.

Summary and Reflection

Now that you have completed the chapter of robot, it's time to reflect on your learning and ensure you have met the goals set for the chapter. Follow these steps to complete the checklist:

- Carefully read through the checklist provided, which outlines the key learning objectives and goals of the robot chapter. For each item on the checklist, evaluate your

own understanding and progress by checking the corresponding box.
- If you feel confident in your understanding and achievement of the goal, check the box; If you believe there are areas where you need further improvement or clarification, leave the box unchecked.

1. **Understanding of Robot Technology:**
   ☐ Have I gained a deeper understanding of the fundamental principles, components, and applications of robots, including the PR2?
   ☐ Can I clearly explain these to others?

2. **Critical Thinking Skills:**
   ☐ How effectively did I apply critical thinking skills to tasks such as skimming, scanning, and matching to extract key information from technical texts?
   ☐ Did I identify relevant information and draw logical conclusions?

3. **Reading Comprehension, Vocabulary, and Language Proficiency:**
   ☐ Have I improved my reading comprehension of cloud-computing-related texts, expanding my vocabulary to include terms?
   ☐ Can I accurately define and use these technical terms related to robotics in context?

4. **Collocating Words and Phrases:**
   ☐ Can I identify and analyze the collocation of words and phrases within the context of robotics to enhance my comprehension of the subject matter?
   ☐ Have I practiced using these collocations in sentences to convey information about cloud computing?

5. **Understanding the Impacts of Robotics Development:**
   ☐ How did I explore the impacts of robotics development on employment, social structure, and human lifestyle?
   ☐ Can I articulate both positive and negative implications of such advancements?

6. **Understanding Chinese Robotics Innovation:**
   ☐ Did I gain a deeper understanding of China's achievements in robot technology through reading articles on Chinese robotics innovation?
   ☐ Can I discuss how these innovations influence the global landscape of robotics development?

7. **Presentation Skills:**

☐ How did I develop my presentation skills by creating and delivering presentations on topics related to robots, using tools like PowerPoint?

☐ Did I effectively organize information, use visual aids, and engage the audience through interactive elements?

8. **Reflection and Critical Analysis:**

☐ How did I reflect on and critically analyze the knowledge I gained about robotics, especially in relation to potential future applications and impacts?

☐ Did I consider different perspectives and evaluate the significance of robotics advancements?

Chapter 5
Electric Vehicle

> **Objectives**
>
> In this chapter, you should be able to:
> - Analyze and discuss the current and future developments in Electric Vehicle(EV) technology.
> - Develop critical thinking, reading, and writing skills by evaluating the feasibility and implications of EVs.
> - Demonstrate the ability to summarize and categorize information from complex texts related to EV technology using techniques such as multiple choice, mind mapping, and matching exercises.
> - Expand vocabulary and language proficiency in an academic context through exercises like glossaries, word choice, and translation tasks related to EVs and its terminology.
> - Engage in collaborative discussions and presentations to share insights and perspectives on EVs and its development, both before and after reading relevant material.

 Before You Read

A. Discussion

Look at the pictures below and discuss with a partner.

1. What brands are the cars in the picture? How are the cars different from a traditional car?

2. What are other famous brands of EVs now? Do you know what their advantages are?

B. Skimming and Scanning

Browse the text and answer the questions below.

1. How many sales did Tesla Model 3 make in January, 2020?
A. 36,700
B. 20,000
C. 21,500
D. 36,000

2. What is the drive range for Wuling Hong Guang MINIEV?
A. 2.9 m
B. 665 km
C. 170 km
D. 177 km

3. What is the starting price for MINI Cooper EV in the U. S.?
A. $20,000
B. $23,245
C. $30,745
D. $37,495

 Text

The Most Popular EV Is No Longer a Tesla Transportation
New EV Made by General Motors and Chinese Manufacturers Underlines the Virtues of Lilliputian Size

1. Tesla sales in China more than doubled in 2020, to nearly $6.6 billion, accounting for 21 percent of the company's booming worldwide total. But Tesla no longer makes China's most

popular EV: Meet the tiny Wuling Hong Guang MINIEV, whose equally microscopic price helped it find 36,700 buyers in January. That easily kneecapped the 21,500 sales of the Model 3 sedan, even as Tesla ramps up production at a new plant in Shanghai.

2. The four-passenger Wuling Hong Guang MINIEV went on sale July 2020, a joint project between Wuling, General Motors(GM) and Chinese state-owned automaker SAIC. The doorstop-shaped city car is just 2.9 meters long, about 0.7 meters shorter than the Chevrolet Spark subcompact that's the smallest car sold in America.

3. With a modest 20 kilowatts (27 horsepower), 85 Nm of peak torque, a 100 kph (62 mph) top speed and 170 km (106 miles) of driving range, the Wuling Hong Guang MINIEV is no match for a Model 3 in space, tech, range or performance. But the Tesla, which starts for around $36,000 in China, can't touch the Wuling Hong Guang MINIEV's price: Just $4,500 (RMB 28,800), or $6,000 (RMB 38,800) with a larger, 13.9 kWh battery.

4. A price more in line with motorcycles has made the Wuling Hong Guang MINIEV a best-seller in what's already the world's biggest market (in total sales) for both cars and electrified vehicles. Among 25.1 million cars sold in China in 2020, 5.4 percent were so-called "New Electric Vehicles", a category that includes EVs, hybrids and plug-in hybrids. Since its introduction in late July 2020, the Wuling Hong Guang MINIEV found 200,000 buyers in its first 200 days, according to Irene Shen, head of communications for GM China.

5. The Wuling Hong Guang MINIEV "is tapping into a large consumer base in China, and makes an EV truly affordable for everyone," Shen said.

6. Shorter than even Japan's famous, penny-pinching Kei cars, the Wuling Hong Guang MINIEV is designed to negotiate tight city streets and tighter parking. Four passengers fit aboard, with room for two 26-inch suitcases or a stroller when rear seats are folded. The Wuling Hong Guang MINIEV can recharge its larger battery in nine hours on China's standard 220 V, 50 Hz outlets.

7. Compared with some bargain-basement cars of the past—most notoriously, India's star-crossed Tata Nano—the Wuling Hong Guang MINIEV seems more engineered to modern standards. GM says 57 percent of the structure is made from high-strength steel. Anti-lock brakes, electronic brake-force distribution, ISOFIX rear child-safety restraints

and ultrasonic rear-parking sensors are standard; but not the electronic stability control that's been required on new cars in the U.S. and EU for several years, or an automated emergency braking system. While GM cites "16 rigorous safety tests" for the Wuling Hong Guang MINIEV, it did not confirm whether the car could meet current crash-test standards in Western nations.

8. Westerners, or Tesla fans, might be tempted to scoff at the Wuling Hong Guang MINIEV's speck-sized body, glacial acceleration or limited range. But while the Wuling Hong Guang MINIEV is too small to crack the size-obsessed American market, its philosophy may be instructive; especially for cities and nations looking to "microtransit" as a potential solution to traffic-clogged streets, emissions and urban quality-of-life issues.

9. The Wuling Hong Guang MINIEV weighs just 665 kilograms (1,466 pounds), which helps it squeeze 170 km of range from just 13.9 kWh of battery.

10. Batteries remain by far the most expensive part of an EV, even as costs for lithium-ion cells have fallen toward $100 per kilowatt hour. Analysts say the battery in a typical electric family sedan can cost between $10,000 and $15,000—roughly double the price of the entire Wuling Hong Guang MINIEV. So the Wuling Hong Guang MINIEV creates a virtuous cycle: The smaller and lighter the car, the less battery it needs to deliver a given range. And a smaller battery itself weighs less. That keeps a lid on mass and costs, for automakers and consumers.

11. America, in fact, has an analog for the Wuling Hong Guang MINIEV: The new electric MINI Cooper SE. In the current EV climate, obsessed with the range race to 400 miles or more, the MINI Cooper SE's 110 mile (177 km) driving range can make it seem as quaint as the Wuling Hong Guang MINIEV. In some quarters, short-hop EVs are seen as a dead-end in design and market viability.

12. The U.K.-based, BMW-owned MINI is a relative giant compared with Wuling's "MINI," including more than twice the curb weight (1,430 kilograms). Accordingly, the MINI Cooper SE carries more than double the battery to deliver a similar official range of 177 kilometers (110 miles), albeit with massively superior power and performance. But in America, the MINI Cooper SE is the Lilliputian, with attendant advantages: Its 32.6 kWh, 220 kilogram pack is one-third the size of Tesla's largest packs, and one-sixth the size of

the brawny, 200 kWh packs that GM will begin stuffing into its longest-range GMC Hummer EV later this year. Setting aside any Muskian advantages in Gigafactory battery costs, the MINI Cooper SE' battery should cost one-third the price of Tesla's.

13. It all leads to the MINI Cooper SE being among the slimmest and most affordable EVs in America. The MINI Cooper SE starts from just $30,745, versus $37,495 for the larger-batteried Chevy Bolt, and $39,125 for a Nissan Leaf Plus. Subtract a $7,500 federal tax credit, and the luxurious, sharp-handling MINI Cooper SE falls to $23,245, on par with gasoline econoboxes like the Toyota Corolla. Incentives in California and other states can push the price closer to $20,000, making the MINI Cooper SE a potential electric steal.

14. Light makes right in another way: In my testing in Miami and New York, the MINI Cooper SE proved it can easily top its official EPA ratings, covering closer to 210 kilometers (130 miles) in real-world drives. Again, that doesn't sound like much for interstate trips. But for everyday errands or school drop-offs—or in cities that may look to tax or even ban combustion-engine cars—130 miles is more than enough for days of local driving. And EV skeptics consistently fail to acknowledge that owners plug in daily or nightly at work or home. They wake up to a car with maximum range fully restored, and never visit a gas station again.

15. MINI Cooper SE executives underline that, for 78 percent of owners, the MINI Cooper SE is the second, third or even fourth car in the household.

16. The point? For Americans whose driveways are littered with gasoline-burning cars, an EV like the MINI Cooper SE—the closest we may ever get to a speck-sized Wuling Hong Guang MINIEV—can still make a whole lot of sense.

 Reading Comprehension

A. Multiple Choice
Choose the best answer for each question.
1. What percentage of Tesla's worldwide total does China contribute?
A. 11%

B. 21%

C. 31%

D. 41%

2. What is the main reason for the popularity of the Wuling Hong Guang MINIEV?

A. Its luxurious interior

B. Its affordable price

C. Its high-end technology

D. Its large size

3. Who is the joint partner of Wuling in the Wuling Hong Guang MINIEV project?

A. Tesla

B. BMW

C. General Motors

D. Toyota

4. What is the price of the MINI Cooper SE after subtracting the federal tax credit?

A. $23,245

B. $30,745

C. $20,000

D. $25,000

5. Which is NOT a stated advantage of the MINI Cooper SE according to the article?

A. Long driving range

B. Affordable price

C. Sharp handling

D. Compact body

B. Mind Map

How many main parts do you think the article is composed of? Fill in the blanks with the details you read from the article.

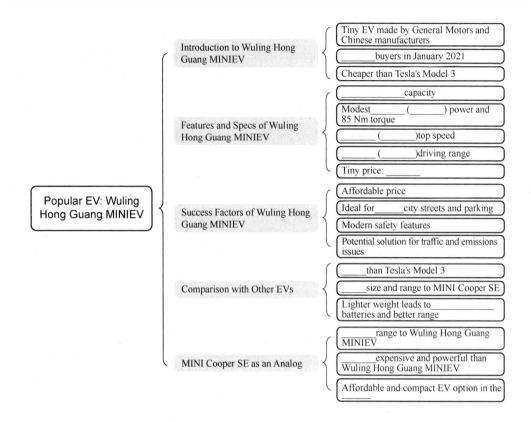

C. Matching

Read the text and decide which paragraph mentions the following information? Write the number of the paragraph before each sentence.

_____ 1. The affordable Wuling Hong Guang MINIEV, priced similarly to motorcycles, has become a top seller in China's booming automotive and electrified vehicle market, attracting 200,000 buyers in its first 200 days since its launch in late July 2020.

_____ 2. The Wuling Hong Guang MINIEV, though unlikely to penetrate the size-driven American market, offers valuable insights, particularly for cities and nations exploring "microtransit" as a means to address congestion, emissions, and urban quality of life challenges.

_____ 3. The Wuling Hong Guang MINIEV embodies a virtuous cycle where its compact and lightweight design significantly reduces the battery requirement, thus minimizing costs for manufacturers and consumers alike, given that batteries are the most expensive component in an EV.

_____ 4. The MINI Cooper SE's lightweight design enables it to exceed its EPA ratings, providing ample range for daily commutes and urban driving, highlighting the convenience of

EVs with daily charging habits and eliminating the need for gas stations.

_____ 5. While it comes equipped with safety features like anti-lock brakes and ISOFIX child restraints, it lacks electronic stability control and automated emergency braking, crucial in Western crash-test standards.

_____ 6. In the US context, the MINI Cooper SE's battery pack is significantly smaller and lighter, making it a cost-effective option despite Tesla's Gigafactory battery cost advantages.

_____ 7. Wuling Hong Guang MINIEV can't match the Model 3 in terms of space, technology, range or performance, but Tesla has no advantage in terms of price.

_____ 8. America's electric MINI Cooper SE offers a comparable analogy to the Wuling Hong Guang MINIEV, challenging the range-focused EV market by showcasing the viability of shorter-range EVs despite prevailing perceptions of their limitations.

D. Cloze

The information below is a summary of the text. Complete the summary by filling in the blanks with the words and phrases provided.

A. affordability and convenience	F. traffic congestion
B. slow acceleration	G. EVs
C. policy incentives	H. compact design
D. outsold	I. urban mobility
E. innovative products	J. low battery consumption

Tesla was once the leader in the Chinese market, but in 2021, the Wuling Hong Guang MINIEV counterattacked and became the best-selling EV in China. Wuling Hong Guang MINIEV quickly occupied the urban travel market with its (1)_____, affordable price and fast charging advantages. Despite falling short of Tesla in terms of technology and range, Wuling Hong Guang MINIEV far (2)_____ Tesla's Model 3 in terms of sales, showing that Chinese consumers are price sensitive and value the convenience of urban use.

Wuling Hong Guang MINIEV has shown a unique competitive advantage in the Chinese market. While some technical specifications are not as good as Western cars, such as (3)_____ and limited range, it has passed rigorous safety tests, met modern engineering standards and had small body size, giving it potential in certain markets and cities. Wuling Hong Guang MINIEV's "microtransit" concept offers a potential solution for cities and countries with severe (4)_____.

At the same time, the characteristics of Wuling Hong Guang MINIEV such as lightweight, (5)_____ and high driving range have promoted the progress of EV technology and provided more choices for the market. Short-range EVs such as the Wuling Hong Guang MINIEV and MINI Cooper SE also show great potential in the U.S. market, and despite limitations in range and performance, their (6)_____ make them ideal for (7)_____.

(8)_____ and the popularity of charging facilities have also created favorable conditions for the development of the EV market. With the continuous progress of EV technology and policy support, it is expected that more people will choose (9)_____ as a mode of travel. The success stories of short-range EVs such as Wuling Hong Guang MINIEV show that (10)_____ adapted to market demand and consumer preferences will stand out in the EV market.

Language Building

A. Glossary

Proper Nouns
Tesla 特斯拉，美国一家电动汽车及能源公司，该公司产销电动汽车、太阳能板、储能设备等。特斯拉努力为每一个普通消费者提供其消费能力范围内的纯电动汽车；特斯拉的愿景是"加速全球向可持续能源转变"。 **General Motors（GM）** 通用汽车公司，成立于 1908 年 9 月，自从威廉·杜兰特创建了美国通用汽车公司以来，通用汽车公司在全球生产和销售别克、雪佛兰、凯迪拉克、GMC 及霍顿等一系列品牌车型并提供服务。2014 年，通用汽车公司旗下多个品牌全系列车型畅销于全球 120 多个国家和地区。 **Wuling Hong Guang MINIEV** 五菱宏光 MINIEV，是上汽通用五菱推出的第一款介于商用车和乘用车的跨界自主研发产品，该产品有流畅的外形设计，多样化、实用性的宽敞驾乘空间。 该产品在动力性和经济性上的完美平衡，以及在操控性和安全性上的实力表现，颠覆了人们对商务车的传统印象。 该产品以其小排量、巧妙设计、精致内饰、强有力的跨级动力，以及 4.48～6.68 万元的售价，在"大块头"云集的商务车市场中开创了一片新天地。"小巧、精细、实用"是业内人士对该产品的一致评价。

SAIC

中华人民共和国国家工商行政管理总局（State AdMINIstration for Industry and Commerce of the People's Republic of China），简称国家工商总局，是中华人民共和国国务院主管市场监督管理和有关行政执法工作的正部级直属机构。

Chevrolet

雪佛兰，也称为Chevy，1918年被通用汽车公司并购，现在为通用汽车公司旗下最为国际化和大众化的品牌。

Hybrids Vehicle

混合动力汽车，是指驱动系统由两个或多个能同时运转的单个驱动系统联合组成的车辆，车辆的行驶功率依据实际的车辆行驶状态由单个驱动系统单独或多个驱动系统共同提供。

通常所说的混合动力汽车，是指油电混合动力汽车（Hybrid Electric Vehicle, HEV），其采用传统的内燃机（柴油机或汽油机）和电动机作为动力源。

Plug-in Hybrid Electric Vehicle

插电式混合动力汽车，简称PHEV，是介于电动汽车与燃油汽车之间的一种新能源汽车，其内部既有传统燃油汽车的发动机、变速器、传动系统、油路、油箱，也有电动汽车的电池、电动机、控制电路，而且电池容量比较大，有充电接口；它综合了电动汽车和混合动力汽车（HEV）的优点，既可实现纯电动、零排放行驶，也能通过混动模式增加车辆的续航里程。

Kei car

所有轻型和微型车辆的统称。Kei car的全称为MPV Kei car，流行于日本汽车文化。Kei car的特点是车身小，车型有轿车、MPV、SUV、两厢车、三厢车。Kei car有很多特点，但是只要车的排量不超过0.66升，长度不超过3.4米，基本上就可以称为Kei car。Kei car的概念最早诞生于20世纪40年代，因为日本国土面积比较小，道路狭窄，所以出现了Kei car这样的小型车。Kei car有一些缺点，比如客舱小，长途驾驶不舒服，动力不足，运载能力低。

Tata Nano

印度塔塔集团最新款轿车，被冠之全球最便宜的汽车称号。其2500美元的价格极具"杀伤力"，从动力到配置，其都节省到了极致。

ISOFIX
一个关于在汽车中安置儿童座椅的新标准,这一标准正在为众多汽车制造商所接受。

MINI Cooper SE
宝马的一个车型,凭借独特的外观、灵巧的操控性能和出色的安全性能赢得了众多年轻人的青睐。

GMC Hummer EV
GMC 旗下皮卡车型。

Gigafactory
电动汽车制造公司特斯拉的超级电池工厂。

Nissan Leaf
一款五座掀背两厢纯电动汽车,配备了先进的由锂离子电池驱动的车辆底盘,其续航里程可达 160 公里以上,可以充分满足消费者在实际生活中的驾驶需求。

Toyota Corolla
丰田卡罗拉,自 2017 年 11 月发布以来,受到全球消费者关注。

EPA
美国国家环境保护局(Environmental Protection Agency),是美国联邦政府下的一个独立行政机构,主要负责维护自然环境和保护人类健康不受环境影响。

Academic Words	
electric (a.)	电的
microscopic (a.)	极小的,微小的
kneecap (v.)	击败
sedan (n.)	轿车
ramp up	增加,斜升
subcompact (n.)	超小型汽车
tap into	接入,开发
penny-pinching (a.)	小气的,便宜的
negotiate (v.)	谈判,磋商,达成协议

stroller (n.)	折叠式婴儿车
bargain (n.)	便宜货，减价品
notoriously (adv.)	声名狼藉地
restraint (n.)	座椅安全带，安全装置
Ultrasonic (a.)	超声的
rear (a.)	后方的，背部的
Rigorous (a.)	严密的，缜密的
tempted (a.)	禁不住的
scoff at	嘲笑
speck (n.)	小颗粒
Glacial (a.)	冰的，慢的
acceleration (n.)	加速能力（车辆）
obsess (v.)	使着迷
philosophy (n.)	理念
instructive (a.)	有启发性的
clogged (a.)	阻塞的；堵住的
lithium-ion (n.)	锂离子
virtuous (a.)	良性的
analog (a.)	类似物
quaint (a.)	奇特有趣的
viability (n.)	生存能力
Lilliputian (a.)	极小的
attendant (a.)	伴随的，随之产生的
brawny (a.)	强壮的
subtract (v.)	减去，扣除
luxurious (a.)	奢侈的，豪华的
incentives (n.)	激励机制
errand (n.)	差事，跑腿
combustion (n.)	燃烧
skeptic (n.)	怀疑论者
consistently (adv.)	一贯地，始终；一致地
executive (n.)	主管，高管
litter (v.)	使（某事物）充满

B. Words and Phrases

Exercises 1 Word Choice

Use the words in the box to finish the sentences.

microscopic	kneecap	ramp	notoriously	scoff
crack	squeeze	virtuous	viability	albeit

1. The vicious circle is thus transformed into a _____ circle.
2. Tumors are _____ hard to kill.
3. He finally agreed, _____ reluctantly, to help us.
4. In one study conducted at an American university, researchers collected _____ germs from footwear.
5. You have to balance the short-term gains against the long-term competitive _____ of the organization.

Exercise 2 Phrases

Match the words provided below with appropriate one in the box.

1. _____ torque
2. _____ test
3. driving _____
4. _____ weight
5. Anti-lock _____

| crash |
| range |
| brake |
| peak |
| curb |

Exercise 3 Sentence Completion

Complete the sentences by filling in the blanks with phrases in the above exercise.

1. RX300 is powered by a 3-liter V6 engine that produces 201 brake horse power and 283 Nm of _____.
2. _____ dummies have saved lives and provided invaluable data on how human bodies react to crashes, but they are designed to represent normal-weight individuals.
3. Sure, in the past EVs have their problems, namely, a limited _____, and very few recharging points, which limited their use.
4. However, its heavy _____ detracts from performance and fuel economy.
5. _____ system is to keep you from skidding on the road and the new cars have them.

Exercise 4　Translation

Translate the sentences by using the words and phrases you have learned in the above two exercises.

1. 山地气候难以预料是<u>人所共知的</u>。
___

2. 我们通过教育和习惯而<u>更有道德</u>。
___

3. 电动汽车<u>续航里程</u>是现阶段研究的一个重要课题。
___

4. 你不能<u>嘲笑</u>别人的信仰。
___

5. 你仍然可以看到它们被当成宠物，<u>尽管</u>这是非法的。
___

C. Collocation

Exercise 1　Modifiers

Find out the adjectives that modify the noun "range" in the article. The first letter has been provided.

d_____ range (para. 3)
l_____ range (para. 8)
g_____ range (para. 10)
o_____ range (para. 12)
l_____ range (para. 12)
m_____ range (para. 14)

Exercise 2　Blank Filling

Scan the text and complete the sentences containing the word "range".

1. With a modest 20 kilowatts (27 horsepower), 85 Nm of peak torque, a 100 kph (62 mph) top speed and 170 km (106 miles) of _____ ...

2. Westerners, or Tesla fans, might be tempted to scoff at the Wuling Hong Guang MINIEV's speck-sized body, glacial acceleration or_____.

3. So the Wuling Hong Guang MINIEV creates a virtuous cycle: The smaller and lighter the car, the less battery it needs to deliver a _____.

4. Accordingly, the MINI Cooper SE carries more than double the battery to deliver

similar _____ of 177 kilometers (110 miles), albeit with massively superior power and performance.

5. ...200 kWh packs that GM will begin stuffing into its _____ GMC Hummer EV later this year.

6. They wake up to a car with _____ fully restored, and never visit a gas station again.

Exercise 3   Translation

Translate the sentences below from Chinese to English using "range" and its collocation in this section.

1. 只在电力驱动下，C-X75 的续航里程是 68 英里。
_____

2. 这只绿海龟只能在有限的范围里游，因为他只有一只鳍。
_____

3. 您最多可以在一个单元格或一个给定的单元格范围中使用两种条件。
_____

4. 最大射程将取决于地形和环境。
_____

5. 他渴望与美国就减少远程核导弹达成协议。
_____

D. Terminology

Exercise 1   Table Filling

Read the article and find the English technical terms according to the Chinese equivalents.

English Technical Terms	Chinese Equivalents
	峰值扭矩
	续航里程
	防抱死系统
	电子刹车力分配系统
	撞击试验
	空车重量

Exercises 2 Blank Filing

Use the terms in the above exercise to complete the sentences below.

1. With a modest 20 kilowatts (27 horsepower), 85 Nm of _____, a 100 kph (62 mph) top speed and 170 km (106 miles) of _____, the Wuling Hong Guang MINIEV is no match for a Model 3 in space, tech, range or performance.

2. _____, _____, ISOFIX rear child-safety restraints and ultrasonic rear-parking sensors are standard;

3. While GM cites "16 rigorous safety tests" for the Wuling Hong Guang MINIEV, it did not confirm whether the car could meet current _____ standards in Western nations.

4. The U.K.-based, BMW-owned MINI is a relative giant compared with Wuling's "MINI," including more than twice the _____ (1,430 kilograms).

Critical Reading and Writing

A. Brainstorming

Work in groups. Fill in the table according to the instruction.

List three aspects about the significance of developing electric vehicles, summarize your ideas and put each in one sentence.

Aspects	Significance
1	1
2	2
3	3

B. Critical Reading

Exercise 1: The following are the answers to the question provided by ERNIE Bot. Read the introduction, and then match the headings and the corresponding content in the body part.

Heading A: Economic Impacts

Heading B: Urban Mobility and Infrastructure

Heading C: Environmental Sustainability

Heading D: Public Health

Heading E: Technological Advancement

The Significance of Developing Electric Vehicles

I. Introduction

The development of electric vehicles (EVs) has become a global priority, driven by the need to address environmental challenges, promote economic growth, and advance technological innovation. EVs offer a range of benefits that span multiple aspects, including environmental sustainability, economic impacts, technological advancement, urban mobility, and public health.

a. _____:

EVs significantly reduce greenhouse gas emissions compared to traditional Internal Combustion Engine Vehicles (ICEVs), helping to mitigate climate change. By eliminating tailpipe emissions, EVs improve air quality in urban areas, reducing respiratory health issues and improving the overall quality of life for residents. This environmental benefit is particularly important given the increasing urgency of addressing climate change and air pollution.

b. _____:

EVs create new job opportunities in the manufacturing, installation, and maintenance of charging infrastructure. This stimulates economic growth and supports the transition to a clean energy economy. In addition, EVs drive innovation in battery technology, electric motors, and other components, further boosting economic growth and competitiveness. Over time, EVs are expected to have lower maintenance costs compared to ICEVs, providing economic benefits to consumers.

c. _____:

EVs push the boundaries of battery energy density, motor efficiency, and vehicle autonomy, driving innovation in related technologies. This technological advancement enables the integration of smart grid technologies, Vehicle-to-Grid (V2G) communication, and self-driving capabilities. These advancements not only improve the performance and efficiency of EVs but also contribute to the development of a more intelligent and interconnected transportation system.

d. _____:

Smaller EV models, such as city cars and scooters, can help reduce congestion in urban areas by encouraging more efficient use of road space. In addition, EVs are quieter compared to ICEVs, improving the urban soundscape and reducing noise pollution. Furthermore, investing in EV infrastructure, such as charging stations, prepares communities for a future where EVs become the norm. This reduces the risk of having to retrofit outdated infrastructure later and enables the smooth transition to a more sustainable transportation system.

　　e. _____:

By reducing air pollution, EVs mitigate respiratory health issues and improve the overall quality of life for residents. In addition, EVs are generally safer in terms of crashworthiness and fire risk compared to some ICEVs, providing a safer option for transportation. This public health benefit is particularly important given the increasing concern about the impact of air pollution on human health.

In conclusion, the development of electric vehicles brings significant benefits in terms of environmental sustainability, economic impacts, technological advancement, urban mobility, and public health. These benefits position EVs as a key solution for addressing the challenges of climate change, air pollution, and sustainable transportation.

Exercise 2: Do you agree with ERNIE Bot on the above ideas? Can you think of more aspects? Share your ideas with the class.

C. Essay Writing

Having engaged in discussions and vocabulary preparation, you've likely generated numerous insightful ideas. Now, it's time to reflect on these ideas and the knowledge you've acquired by crafting an essay. Use the following instructions to guide your writing process:

Topic

The Significance of Developing EVs in … (a specific aspect)

Background Information

We live in a time when technological advances are reshaping the way we live and work. One of the most transformative trends in this revolution is the rise of EVs. These vehicles, powered by electricity rather than fossil fuels, will revolutionize the automotive industry and bring profound changes to every aspect of our lives.

Whether environmental, technological, economic or social, EVs play a role in many ways. So, in which aspect has the vigorous development of EVs brought us or will bring us what kind of changes? What is the profound significance of developing EVs for us?

Instructions

In an essay of approximately 300-450 words, <u>choose an aspect and analyze the significance of developing EVs in that aspect</u>. Provide specific evidence and arguments to support your opinion.

Your essay should include the following components:

Introduction (approximately 50-75 words): Briefly introduce the topic and provide context for your analysis.

- State your thesis or main argument about the significance of developing EVs in the chosen aspect.

Body Paragraphs (approximately 250-350 words): Present your argument in detail, supported by specific evidence and examples in 3 separate paragraphs.

- Describe the particular aspect of your choice in detail, explaining why it is important and why it is relevant to the development of EVs.
- Delve into how EVs can have an impact in that particular area.
- Use specific examples or case studies to support your analysis.
- The focus should be on the positive.

Conclusion (approximately 50-75 words): Summarize your main points and restate your thesis in light of the evidence presented.

- Emphasize the need to continue to develop and promote EVs in the future.
- Make appeals or suggestions to encourage the reader or the government to take further action.

* Ensure that your essay is well-structured, logically organized, and supported by evidence from reputable sources. Use clear and concise language, and proofread your work carefully for grammar, punctuation, and spelling errors.

Discussion and Presentation

A. Group Discussion

Exercise 1: Think and discuss the questions below.

1. What role do EVs play in China's energy transition away from fossil fuels, balancing energy security and environmental protection, and what are the challenges and opportunities in this process?

2. What efforts have Chinese EV manufacturers made in technological innovation, especially in areas like autonomous driving and intelligent connectivity, and how do they compete with international players?

3. What are the current trends and opportunities for international cooperation in China's

EV industry, especially in terms of technology transfer and market access, and how does this affect the industry's global competitiveness?

4. How do EVs contribute to China's economic growth, job creation, and industrial upgrading, and what role do they play in China's economic structural transformation?

Exercise 2: Read the article below. Search any information related to the above questions. Do you think the answers provided in the article match yours? Share your ideas with the class.

B. Extended Reading-EV in China

Read the article, highlight any ideas or language that will help you in your presentation in the next session.

Electric Passenger Vehicles, Lithium Batteries and Solar Cells Are Hot in Canton Fair

On October 18th, 2023, the third Belt and Road Forum for International Cooperation (BRF) High-level Forum on Green Development was held grandly in Beijing for worldwide green, low-carbon and sustainable development. At the same time, at the 134th Canton Fair, new energy projects like electric passenger vehicles, lithium batteries and solar cells can be seen everywhere, attracting overseas buyers from countries along the Belt and Road.

Electric passenger vehicles, lithium batteries and solar cells have become the new business signature for China's foreign trade exports. At this year's Canton Fair, these three kinds of products from China are not only popular in countries along the Belt and Road but also driving the green transformation of industries in these countries.

In front of Skyworth's booth, Pu, a buyer from Africa, was consulting about the photovoltaic solutions and energy storage products shown by Skyworth this year. "These new energy technologies are very important to us. We need some photovoltaic solutions that can reach the end users and some commercial and industrial energy storage products." Pu excitedly told the reporter, "We found that there are too many new energy products on site. There are still so many products to be known after attending four days!"

At the booth of Leoch International, an Ethiopian buyer who attended the Canton Fair for the first time has just reached a cooperation intention with Leoch regarding solar cells. He told reporters that there is a lack of imported resources for solar cells in his local area. He is looking forward to attending the Canton Fair and is willing to continue cooperating with Chinese companies.

From the Canton Fair, it is not difficult to see that electric passenger vehicles, lithium batteries and solar cells are popular among countries along the Belt and Road. Data shows that in the first three quarters of 2023, the total export of these three kinds of products reached 798.99 billion yuan, a year-on-year increase of 41.7%, accounting for 4.5% of China's exports with an increase of 1.3 percentage points. The export value has maintained in double figures for 14 consecutive quarters.

In recent years, as green and low-carbon have become a trend worldwide, the Chinese new energy industry has ushered in huge market dividends. According to statistics, nearly 90% of the global photovoltaic industry's production capacity is in China. In 2022, China's photovoltaic product export scale exceeded 50 billion dollars for the first time, a year-on-year increase of 80.3%.

The upgrade of typical export products from clothing, furniture and home appliances to electric passenger vehicles, lithium batteries and solar cells reflects the fast transformation and upgrading of China's advanced manufacturing and energy industries.

From a 3.5kW portable EV charger to a 720kW split liquid-cooled EV fast charger, and then to the EV chargers with solar panels, Loncin Motor has brought a series of new energy products to this Canton Fair after its strategic transformation. It is reported that in April 2023, Loncin has launched a batch of new energy products targeting the European market.

The new energy battery products exhibited by Guangdong JYC Battery in 2023 have incorporated charging network technology from other industries based on the previous generation. Innovative manufacturing processes have resulted in at least a 30% improvement in the cycle life, high current discharge performance, and self-discharge of the battery products.

It is reported that, the participating companies of this year's Canton Fair have uploaded about 430,000 green and low-carbon products and about 110,000 intelligent products, demonstrating the vigorous development of the Canton Fair and the continuous optimization of China's foreign trade structure.

C. Presentation

After you have read the article, please choose one of the following topics to develop your ideas. Make a presentation with PowerPoint to the class.

Topic 1　International Cooperation and Export Trends of EVs under the Belt and Road Initiative
- Introduce the connection between the Belt and Road Initiative and the EV industry.
- Analyze the market demand and growth trends for EVs in countries along the Belt and Road.
- Discuss how Chinese EV enterprises expand their international market through platforms such as the Canton Fair.
- Explore the positive impact of EV exports on China's economy, energy structure, and environmental protection.
- Predict the future development potential and challenges of EVs in countries along the Belt and Road.

Topic 2　Technological Innovation and Industrial Upgrading of EVs
- Introduce the latest development trends in EV technology.
- Analyze the efforts and achievements of Chinese EV enterprises in technological innovation.
- Discuss how technological innovation in EVs promotes the transformation and upgrading of China's manufacturing and energy industries.
- Provide examples of how technological innovation enhances the market competitiveness of EVs and meets consumer demand for green, efficient, and intelligent transportation.
- Explore the importance of technological innovation in EVs for sustainable development and addressing climate change.

Note:
- Craft a visually appealing PowerPoint with appropriate colors and images.
- Keep each slide concise, using fewer than 10 words.
- Use Pictures, illustrations or forms to make your point.
- Emphasize positive concepts and messages throughout your presentation.
- Aim to deliver your presentation smoothly, without relying on notes, within a timeframe of 4-5 minutes.

Chapter 5 Electric Vehicle

Video

A. Before You Watch

Read out the words below. Choose a word in the box to form an appropriate expression.

linear   motor   combustion   electro   polarity   force   direct electrical   current

- internal _____ engine
- explosive _____
- _____ motion
- _____ magnet
- _____ wires
- magnetic _____
- electric _____
- _____ current
- alternating _____

B. While You Watch

Exercise 1: Discussion

Do you agree that EVs are less likely to fail or require expensive time-consuming maintenance?

Exercise 2: Dictation

Fill in the blanks with words and expressions you have heard from the video.

Surprisingly, the operating principle behind most modern EVs predates the (1) _____ by a number of decades. In 1834, a Dutch professor named Sibranda Stratting of the Netherlands, built his own (2) _____. The catch being its battery was non (3) _____. Internal combustion engine regions work on the principle that fuel and air, when compressed and ignited, cause a tiny explosion. That's the combustion part. This (4) _____ pushes a piston that pistons (5) _____, in concert with a team of fellow pistons, transforms into rotary motion via a mechanical crank shaft. This in turn spins your wheels along the highway.

Conversely, the fundamental principle that drives EVs is (6) _____. Everybody knows how opposing poles on a magnet attract and how a like poles repel each other. So let's imagine an experiment using two magnets, one fixed, the other mounted on a nearby rotating shaft. If the two poles nearest to each other on both magnets share the (7) _____, say north to north. When it on, the shaft will be repelled. Because it's attached to a shaft, the shaft will turn, that is, until the south pole on the shaft magnet is aligned with the north pole on the fixed magnet, whereupon the shaft will again be still. In our imaginary experiment, we've made the shaft turn a half rotation. All very well, but that won't get us very far on the morning commute. Here's where (8) _____ enters the chant. In a fixed or permanent magnet, like the kind you have on your fridge at home, those magnetic poles are rigid and never change. North is always north. South is always south. On an electro magnet, however, which is essentially a core of metal called an (9) _____, this magnetic polarity can be reversed. Imagine one of our experimental magnets is now an electro magnet. If the south pole quickly flips over to north, the fixed magnet will yet again repel the moving magnet, rotating our shaft another half spin. That's a whole spin now. We're slowly getting for a basic illustration of (10) _____, imagine a very simple circuit involving a battery and a light bulb. Electrons flow in one direction from the battery, through the wires, to the light bulb, and back again to the battery. If we remove our battery from the circuit, flip it 180 degrees, then replace it in the circuit, those electrons will still flow around the circuit just in the (11) _____. Either way, the bulb lights up. Electro Magnets, like light bulbs, work whichever direction the electrons are flowing. But rather brilliantly, the polarity of the magnet gets reversed with the flow of electrons. So, to keep our magnets in permanent repel mode, we just need to keep (12) _____ of the magnet. How do we do that?

    One way would be to keep popping out the battery and flipping it around, but that's a lot, a trip to the mechanics with your EV for the sake of a few feet of ground covered. So, the real trick to making our magnet spin, which is essentially how (13) _____ work, is through the so called "inverter". The inverter module on the EV draws (14) _____ from the car battery, and through a clever conference, the nation of quick switches, slick circuitry and capacitors, flips the flow of electrons back and forth nearly 60 times a second. Domestic electric motors, like the one you have in your hair dryer, don't require an inverter. Why? Because the current that comes from your wall outlet is already flipping back and forth. That's why it's named (15) _____ or AC. Batteries of any type can only ever produce DC or direct current. So spinning magnets driven by alternating current passing through coils of wire is essentially what drives EVs. Electric power trains have a number of advantages over the internal combustion engine. For starters, the motion produced by the motor is already rotary in nature. Dirty pistons on an ICE require a complicated, breakable crank shaft just in order to turn their linear motion into rotary movement. So EVs are less likely to fail or require expensive time-consuming maintenance.

Summary and Reflection

Now that you have completed the chapter of electric vehicle, it's time to reflect on your learning and ensure you have met the goals set for the chapter. Follow these steps to complete the checklist:

- Carefully read through the checklist provided, which outlines the key learning objectives and goals of the electric vehicle chapter. For each item on the checklist, evaluate your own understanding and progress by checking the corresponding box.
- If you feel confident in your understanding and achievement of the goal, check the box; If you believe there are areas where you need further improvement or clarification, leave the box unchecked.

1. **Understanding of EVs:**
   ☐ Can I explain the advantages of EVs, the challenges of future development and what is the significance of developing EVs?
2. **Critical Thinking Skills:**
   ☐ How effectively did I apply critical thinking skills to tasks such as skimming, scanning, and matching to extract key information from technical texts?
   ☐ Did I identify relevant information and draw logical conclusions?
3. **Reading Comprehension, Vocabulary, and Language Proficiency:**
   ☐ Have I improved my reading comprehension of EV-related texts, expanding my vocabulary to include terms?
   ☐ Can I accurately define and use these technical terms in appropriate contexts?
4. **Collocating Words and Phrases:**
   ☐ Can I identify and analyze the collocation of EV-specific words and phrases to enhance my comprehension of the subject matter?
   ☐ Have I practiced using these collocations in sentences to convey information about EVs?

5. **Understanding of Impacts:**

☐ How did I explore the environmental and societal impacts of EVs, considering factors like carbon emissions reduction, energy efficiency, and economic impacts?

☐ Can I articulate the broader implications of such actions?

6. **Presentation Skills:**

☐ How did I develop my presentation skills by creating and delivering presentations on topics related to EVs, using tools like PowerPoint?

☐ Did I effectively organize information, use visual aids, and engage the audience through interactive elements?

7. **Reflection and Critical Analysis:**

☐ How can I reflect on and critically analyze the knowledge gained throughout the chapter, especially regarding the significance of the development of EVs and the development and innovation of EV technology in China?

☐ Did I consider different perspectives on EVs, evaluating their significance for the environment, energy, economic and society at large?

Chapter 6
Aerospace

Objectives

In this chapter, you should be able to:
- Recognize the significance of reusable rockets in space exploration.
- Extract relevant information and identify main ideas and details.
- Expand vocabulary related to aerospace engineering.
- Identify and define technical terms associated with aerospace technology.
- Apply critical thinking skills to analyze solutions for space debris removal.
- Develop ideas for enhancing international collaboration in space exploration.
- Reflect on extended reading material regarding the living space on China's space station.
- Create and deliver presentations on chosen topics related to space exploration.
- Engage in pre-watching activities by selecting appropriate words and phrases.
- Participate in discussions regarding rocket staging phases during rocket launching.

 Before You Read

A. Discussion

Look at the pictures below and discuss with a partner.

1. Why do rocket stages need to be dumped during rocket launching?

2. What is the significance of a reusable rocket?

B. Skimming and Scanning

Browse the text and answer the questions below.

1. How does China ensure the safety of residents in rocket launch areas?

A. Calculate the areas in which stages and side boosters fall

B. Build all the spaceports on the coast

C. Issue local warnings in advance

D. Implement evacuation orders in advance

2. Which rockets were launched on the coast?

A. Long March 5B

B. Long March 7A

C. Long March 8

D. Long March 3B

3. What did Long March 7A use?

A. relatively clean kerosene

B. liquid oxygen propellant

C. Toxic fuel

D. Oxidizer mix

Text

China Tries To Solve Its Rocket Debris Problem
Warnings Aren't Enough, But Reusable Rockets Might Be

1. On the morning of 17 June, 2021, China launched the first astronauts to its Tianhe space station module, with a Long March 2F rocket sending the crewed Shenzhou-12 spacecraft into orbit.

2. Overlooked from this major success, however, was that downrange from its Gobi Desert

launch site, empty rocket stages fell to ground. As getting to orbit is about reaching a velocity high enough to overcome gravity (roughly 7.8 kilometers per second), rockets consist of stages, with tanks and engines optimized for flying through the atmosphere dumped to reduce mass once they empty. A video posted on the Twitter-like Sina Weibo emerged that seems to show one of these as part of the recovery process, with apparent residual, hazardous propellant leaking from the broken boosters. According to the source, road closures and evacuations allowed a safe clean up.

3. However, where the stages fall and how they are dealt with is not as haphazard as it appears. The areas in which stages and side boosters fall are calculated to avoid major populations, and local warnings and evacuation orders are issued and implemented in advance. The array of amateur footage of falling rocket stages backs up the claim that the events are known and expected. No casualties from these events have so far been reported.

4. Yet this approach is costly, disruptive and not without continued risks and occasional damage.

5. To mitigate and eventually solve the problem, the China Aerospace Science and Technology Corporation(CASC), the country's main space contractor is developing controllable parafoils to constrain the areas in which the stages fall, most recently tested on a Long March 3B launch from Xichang, southwest China, one of China's workhorse launchers, which most frequently threaten inhabited areas. Grid fins, the kind which help guide Falcon 9 rocket core stages to landing areas, have been tested on smaller Long March 2C and Long March 4B launch vehicles.

6. The latter step is part of attempts to develop rockets that can launch, land and be reused like the Falcon 9, thus controlling the fall of large first stages.

7. The Long March 8, which had a debut (expendable) flight in December, 2020, is expected to be CASC's first such launcher. Chinese commercial companies including Landspace, iSpace, Galactic Energy and Deep Blue Aerospace, are also working on reusable rockets. Low altitude hop tests are expected this year. Though, as SpaceX has demonstrated in a self-depreciating compilation, landing rockets vertically is no mean feat.

8. Additionally, the Long March 7A rocket, which launches from the coast, is expected to

eventually replace the aging Long March 3B rocket for launches to geostationary orbit. As a bonus the Long March 7A uses relatively clean kerosene and liquid oxygen propellant instead of the toxic fuel and oxidizer mix used by the Long March 3B. The Long March 2F used for crewed launches could also be replaced by a Long March 7 or a low-Earth orbit version of a new-generation rocket being developed to eventually send Chinese astronauts to the moon.

9. China has also developed the ability to conduct launches at sea. Its new Long March 5B, which launches from the coast, also has its own particular issue, namely the first stage actually (and unusually) reaching orbit, and returning to Earth wherever and whenever the atmosphere drags it back down.

10. Despite these measures, launching over land and population centers is fraught with risk. The debris issues and events above are when things go well; failures could bring yet greater danger. China's space industry and activities have expanded greatly in recent decades, including lunar and planetary exploration, human spaceflight, remote sensing, spy, weather and other satellites, as well as a new commercial space sector.

Reading Comprehension

A. Multiple Choice

Choose the best answer for each question.

1. What event occurred on the morning of June 17th?
   A. Launch of the Tianhe space station module
   B. Launch of the Shenzhou-12 spacecraft
   C. Recovery of empty rocket stages
   D. Development of controllable parafoils

2. What is a major concern regarding China's space launches?
   A. Lack of security measures
   B. Overuse of rocket stages
   C. Frequent accidents and casualties
   D. Falling rocket stages near populated areas

3. What approach has China taken to mitigate the problem of falling rocket stages?

A. Issuing evacuation orders post-launch

B. Avoiding inhabited areas for rocket stages to fall

C. Implementing grid fins on launch vehicles

D. Developing reusable rockets like the Falcon 9

4. What is the purpose of controllable parafoils in China's space program?

A. To guide rocket stages during launch

B. To reduce the cost of rocket launches

C. To restrict the areas where rocket stages fall

D. To increase the speed of rocket reentry

5. Which rocket is expected to replace the aging Long March 3B rocket for launches to geostationary orbit?

A. Long March 2F

B. Long March 8

C. Long March 5B

D. Long March 7A

B. Mind Map

How many main parts do you think the article is composed of? Group the paragraphs and fill in the blanks with the information you read from the article.

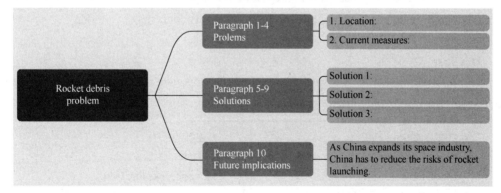

C. Matching

Read the text and decide which paragraph mentions the following information? Write the number of the paragraph before each sentence.

_____ 1. Despite the calculated efforts to avoid populated areas, there have been instances of rocket stages falling near schools and inhabited regions.

_____ 2. Furthermore, Chinese commercial companies like Landspace and iSpace are also exploring reusable rocket technologies.

_____ 3. China Aerospace Science and Technology Corporation(CASC) is actively developing technologies like controllable parafoils and grid fins to address the rocket debris issue.

_____ 4. The historical context of China's spaceports reveals that most were built from 1947 to 1991, prioritizing security over proximity to coastlines.

_____ 5. Additionally, advancements in sea-launched rockets like the Long March 5B offer alternative solutions to land-based launches, albeit with their unique challenges.

_____ 6. Despite ongoing efforts to manage space debris, concerns persist over the risks associated with rocket launches, highlighting the need for continuous innovation and risk mitigation strategies.

_____ 7. China's Long March 7A rocket, using cleaner propellants, is poised to replace older models and reduce environmental impact.

_____ 8. In a recent space launch by China, empty rocket stages descended from the sky, prompting concerns about hazardous propellant leakage upon impact.

_____ 9. These innovations aim to guide rocket stages to designated landing areas, akin to SpaceX's Falcon 9 recovery methods.

_____ 10. The Long March 8 rocket, anticipated to be China's first reusable launcher, signifies a significant step towards mitigating space debris concerns.

D. Cloze

The information below is a summary of the text. Complete the summary by filling in the blanks with the words and phrases provided.

A. empty rocket stages	F. populated
B. space	G. risks
C. space exploration	H. space exploration
D. technologies	I. innovation
E. inhabited	J. debris

The article discusses China's recent achievements in (1) _____, notably the launch of the Tianhe space station module and the crewed Shenzhou-12 spacecraft. However, it also highlights a concerning issue regarding (2) _____ falling to the ground after launch, particularly in areas near (3) _____ regions. Despite the calculated efforts to minimize (4) _____ and implement safety measures, these incidents remain a challenge for Chinese (5) _____ launches. To address this problem, the China Aerospace Science and Technology Corporation(CASC) is developing (6) _____ such as controllable parafoils and grid fins to guide rocket stages to designated landing areas. Additionally, efforts are underway to develop reusable rockets, akin to the Falcon 9, to control the fall of large first stages and mitigate (7) _____ risks. Despite these advancements, concerns persist over the potential hazards associated with rocket launches near (8) _____ areas. The article underscores the need for ongoing (9) _____ and risk management in China's space industry to ensure the safety and success of future missions amidst its expanding (10) _____ endeavors.

Language Building

A. Glossary

Proper Nouns
Tianhe Space Station 　　天和空间站,一般指天和核心舱。天和核心舱是中国空间站天宫的组成部分。2021年4月29日11时23分,长征五号B遥二运载火箭搭载天和核心舱,在中国文昌航天发射场发射升空。 **Long March 2F rocket** 　　长征二号F火箭,是1999年中国发射神舟载人飞船的两级捆绑助推器运载火箭。 **Gobi Desert** 　　戈壁沙漠,是世界上巨大的荒漠与半荒漠地区之一,绵亘在中亚浩瀚的戈壁大地上,跨越蒙古国和中国广袤的空间。

Wenchang Launch Center
文昌航天发射场，位于海南省文昌市龙楼镇，隶属于西昌卫星发射中心，是中国首个开放性滨海航天发射基地，也是世界上为数不多的低纬度发射场之一。

China Aerospace Science and Technology Corporation (CASC)
中国航天科技集团有限公司（简称"航天科技"或"中国航天"），是在中国战略高技术领域拥有自主知识产权和著名品牌、创新能力突出、核心竞争力强的国有特大型高科技企业，成立于1999年7月1日。

Grid fins
栅格翼，是一种较少采用的气动面形式，由众多薄的栅格壁镶嵌在边框内形成。栅格壁在边框内的布局形式是多样的，基本的有两种，一种是框架式，一种是蜂窝式。

Falcon 9 rocket
猎鹰9号火箭，是美国SpaceX公司研制的可回收式中型运载火箭，于2010年6月4日完成首次发射，于2015年12月21日完成首次回收。

Academic Words	
module (n.)	（航天器的）舱
crewed (a.)	载人的
downrange (a.)	沿试验航向的
stage (n.)	（火箭的）级
velocity (n.)	速度（矢量）
gravity (n.)	重力，地心引力
roughly (adv.)	粗略地，大约
tank (n.)	（储存液体或气体的）箱，罐，缸
optimize (v.)	优化，充分利用
dump (v.)	丢弃，扔掉
mass (n.)	[力] 质量
emerge (v.)	浮现，出现
residual (a.)	残留的；（数量）剩余的
hazardous (a.)	危险的，有害的
propellant (n.)	推进燃料
booster (n.)	助推器，推进器
closure (n.)	（尤指通路、边境的）封锁

evacuation (n.)	撤离，疏散
fledging (a.)	刚起步的
paramount (a.)	至为重要的，首要的
preemptive (a.)	先发制人的
haphazard (a.)	偶然的，无计划的
issue (v.)	发表，颁布
implement (v.)	执行，贯彻
array (n.)	一系列，大量
amateur (a.)	非专业的，业余爱好的
footage (n.)	一组（电影，电视）镜头
disruptive (a.)	引起混乱的，破坏的
mitigate (v.)	［正式］减轻，缓和
contractor (n.)	承包商，立约人
parafoil (n.)	翼伞
constrain (v.)	限制，约束
inhabit (v.)	居住于，栖居在
debut (n.)	首次登台，（新事物的）问世
altitude (n.)	海拔高度
self-depreciating (a.)	自贬的；谦虚的
compilation (n.)	编纂，汇编
vertically (adv.)	垂直地
geostationary (a.)	与地球旋转同步的
kerosene (n.)	煤油，火油
toxic (a.)	有毒的，引起中毒的
oxidizer (n.)	［助剂］氧化剂
fraught (a.)	充满（糟糕或令人讨厌的事物）
lunar (a.)	月亮的，月球的
planetary (a.)	行星的

B. Words and Phrases

Exercises 1   Word Choice

Use the words in the box to finish the sentences.

optimize	issue	mitigate	haphazard	disruptive
debut	toxic	implement	evacuation	vertically

1. The police have _____ an appeal for witnesses.
2. They might eat something _____ and damage their health.
3. You can move the camera both _____ and horizontally.
4. Exercise to some degree could _____ the psychological stress reactions.
5. The economy is in danger of collapse unless far reaching reforms are_____.

Exercise 2   Phrases
Match the words provided below with appropriate one in the box.

1. remote	flight
2. overcome	Gravity
3. toxic	the problem
4. debut	Fuel
5. mitigate	Sensing

Exercise 3   Sentence Completion
Complete the sentences by filling in the blanks with phrases in the above exercise.

1. President George W. Bush ordered the destruction of the satellite, which was carrying a _____ called hydrazine.
2. A plane is heavy and needs a lot of thrust to produce enough lift to _____ and take off.
3. When odd things break or don't work, you just learn the (sometimes arcane) rules to _____ and move on.
4. _____ is the use of satellites to search for and collect information about the earth.
5. The world's newest long-haul budget airline got off to an inauspicious start yesterday after its _____ from Hong Kong to London was cancelled because it lacked permission to fly over Russia.

Exercise 4   Translation
Translate the sentences by using the words and phrases you have learned in the above two exercises.

1. 王宫发言人刚发布了一项声明。

2. 新算法优化了 Google 广告位排列顺序，大大增加了广告效果。

3. 运动能在一定程度上减轻不良心理应激反应。
___

4. 他昨天首次出战篮球联赛。
___

5. 疏散工作确实看起来缺乏计划。
___

C. Collocation

Exercise 1   Modifiers
Complete the first words in the following phrases by reading in the text.

h_____propellant(para. 2)
p_____strikes (para. 3)
f_____nuclear weapon capabilities (para. 3)
d_____flight (para. 7)
r_____rockets (para. 7)
t_____fuel (para. 8)

Exercise 2   Blank Filling
Scan the text and complete the sentences with the collocations in the above exercise.

1. As a bonus the Long March 7A uses relatively clean kerosene and liquid oxygen propellant instead of the _____ and oxidizer mix used by the Long March 3B.

2. The Long March 8, which had a _____ in December, is expected to be CASC's first such launcher.

3. A video posted on the Twitter-like Sina Weibo emerged that seems to show one of these as part of the recovery process, with apparent residual, _____ leaking from the _____.

4. China's three main spaceports were built from 1947 to 1991 when security was paramount, with the U.S. and Soviet Union considering the possibility of _____ on facilities linked to China's _____.

Exercise 3   Translation
Translate the sentences below from Chinese to English using the adjective below.

hazardous	fledging	preemptive
debut	reusable	toxic

1. 他们没有办法处理掉自己产生的有害废料。

2. Paula 是唯一的首张唱片 4 次荣登排行榜榜首的艺人。

3. 该技术还处于起步阶段。

4. 能从其他废料中分离出可重复利用成分是很关键的。

5. 有时你需要先发制人。

D. Terminology

Exercise 1   Table Filling

Read the article and find the English technical terms according to the Chinese equivalents.

English Technical Terms	Chinese Equivalents
	侧置助推器
	低空跳跃试验
	对地静止轨道
	液氧推进剂
	月亮和行星际探索
	遥感

Exercises 2   Blank Filling

Use the terms in the above exercise to complete the sentences below.

1. China's space industry and activities have expanded greatly in recent decades, including _____, human spaceflight, _____, spy, weather and other satellites, as well as a new commercial space sector.

2. _____ are expected this year.

3. As a bonus the Long March 7A uses relatively clean kerosene and _____ _____ instead of the toxic fuel and oxidizer mix used by the Long March 3B.

4. The areas in which stages and _____ fall are calculated to avoid major populations, and local warnings and evacuation orders are issued and implemented in advance.

5. Additionally, the Long March 7A rocket, which launches from the coast, is expected to eventually replace the aging Long March 3B rocket for launches to _____.

Critical Reading and Writing

A. Brainstorming

Work in groups. Fill in the table according to the instruction.

What innovative methods could be developed to safely remove existing space debris from Earth's orbit? List at least three methods.

Method 1	
Method 2	
Method 3	

B. Critical Reading

Exercise 1: The following are the answers to the question provided by ChatGPT. Which one do you think is the most effective? Number 1(most effective) to 5 (least effective) to the items in the table. Use any language tools to help you find out the meanings of the words and expressions.

⑤ ChatGPT

____ a. Laser ablation:
Utilizing ground-based or space-based lasers to target and vaporize smaller debris, reducing their size and mitigating collision risks.

____ b. Robotic capture:
Deploying autonomous or remotely controlled robots equipped with nets or harpoons to capture and collect large debris items, which can then be brought down to Earth safely or moved to a "graveyard" orbit.

____ c. Electrodynamic tethers:
Attaching conductive tethers to debris, using Earth's magnetic field to create drag and gradually lower the orbit of debris until it burns up in the atmosphere.

____ d. Air-breathing ion thrusters:
Developing ion thrusters that can operate in the very thin upper atmosphere to extend their operational life and be used for nudging debris into deorbiting trajectories.

Exercise 2: Which method(s) mentioned above require(s) international collaboration? How can international collaboration be enhanced to establish and enforce global regulations on minimizing space debris from future launches?

C. Essay Writing

Having engaged in discussions and vocabulary preparation, you've likely generated numerous insightful ideas. Now, it's time to reflect on these ideas and the knowledge you've acquired by crafting an essay. Use the following instructions to guide your writing process:

Topic

Enhancing International Collaboration for Space Debris Regulation.

Background Information

Space debris, also known as space junk or orbital debris, refers to defunct human-made objects orbiting Earth that no longer serve any useful purpose. These objects include discarded rocket stages, old satellites, fragments from spacecraft collisions, and other debris generated from various space missions and activities. Over the years, the accumulation of space debris has become a growing concern due to its potential to pose risks to operational satellites, spacecraft, and even crewed missions.

The proliferation of space debris is attributed to several factors, including historical space missions, satellite launches, and accidental collisions between orbiting objects. As space activities continue to expand, particularly with the emergence of commercial space ventures and the increasing deployment of satellites for telecommunications, navigation, and Earth observation purposes, the risk of collisions and the generation of additional debris also rises.

Space debris poses significant challenges to space operations and safety. Collisions between debris and operational satellites or spacecraft can result in damage or destruction, jeopardizing critical infrastructure and disrupting essential services such as communications, weather forecasting, and scientific research. Furthermore, the sheer abundance of debris in certain orbital regions poses hazards to future space missions, including crewed exploration endeavors and the deployment of new satellites.

Addressing the issue of space debris requires a coordinated and concerted effort at the international level. While individual countries have implemented measures to mitigate debris generation and minimize collision risks, such as end-of-life disposal guidelines and collision avoidance maneuvers, a comprehensive and globally coordinated approach is necessary to effectively manage space debris and ensure the long-term sustainability of space activities.

International collaboration plays a crucial role in establishing and enforcing global regulations on minimizing space debris from future launches. By working together, space-faring nations can develop common standards, share best practices, and implement collective measures to mitigate the generation of debris, track existing objects, and remove hazardous debris from orbit. Enhanced collaboration fosters transparency, trust, and cooperation among nations, laying the foundation for a safer and more sustainable space environment.

Instructions

In an essay of approximately 300-450 words, <u>analyze which industry could be mostly influenced by the advent of the aerospace technology</u>. Provide specific evidence and arguments to support your opinion.

Your essay should include the following components:

Introduction (approximately 50-75 words): Briefly introduce the topic and provide context for your analysis.
- <u>Provide a brief overview of the growing problem of space debris and its implications for space exploration and satellite operations.</u>
- <u>Introduce the importance of international collaboration in addressing this issue and outline the main points to be discussed in the essay.</u>

Body Paragraphs (approximately 250-350 words): Present your argument in detail, supported by specific evidence and examples in 3 separate paragraphs.
- <u>Discuss the role of organizations such as the United Nations Committee on the Peaceful Uses of Outer Space (UNCOPUOS) and the Inter-Agency Space Debris Coordination Committee (IADC) in facilitating international cooperation.</u>
- <u>Explore how diplomatic initiatives can lead to the establishment of binding agreements and treaties aimed at reducing space debris.</u>
- <u>Highlight the importance of fostering trust and mutual understanding among nations to foster long-term cooperation.</u>

Conclusion (approximately 50-75 words): Summarize your main points and restate your thesis in light of the evidence presented.

> - Emphasize the importance of enhanced international collaboration in addressing the space debris problem.
> - Highlight the potential benefits of collective action and the need for sustained commitment from all stakeholders to ensure a safe and sustainable space environment for future generations.
>
> * Ensure that your essay is well-structured, logically organized, and supported by evidence from reputable sources. Use clear and concise language, and proofread your work carefully for grammar, punctuation, and spelling errors.

 ## Discussion and Presentation

A. Group Discussion

Exercise 1: Think and discuss the questions below.

1. What do you think are some of the challenges astronauts face while living and working in space, and how might advancements in space station design address these challenges?

2. How important do you think it is for countries to invest in space exploration and the development of space stations? What potential benefits do you foresee from such investments?

3. Considering the advancements mentioned in the article, what are some potential implications for future space missions and humanity's exploration of space?

Exercise 2: Read the article below. Search for any information related to the above questions. Do you think the answers provided in the article match yours? Share your ideas with the class.

B. Extended Reading-Aerospace in China

Read the article, highlight any ideas or language that will help you in your presentation in the next session.

China Space Station: What Will Astronauts' Living Space Look Like?

The astronauts who will conduct tasks on China's space station will be enjoying a larger space than their predecessors.

The core module Tianhe launched is the first module of China's space station sent to the space. It is not only a control center but a main space where astronauts will live and conduct scientific works.

The airtight cabin of the core module Tianhe offers 50 cubic meters of space for the astronauts, over three times bigger than Tiangong-1 or Tiangong-2 space lab with only about 15 cubic meters, said Bo Linhou, deputy chief designer of the China's space station.

"It's a substantial leap in terms of space for astronauts moving around," Bo said.

The space station will be T-shaped with the core module at the center and two lab capsules separately on each side when the construction is complete. The three-module station can accommodate astronauts with over 100 cubic meters space for living and working, six times bigger than Tiangong-2 space lab.

Six zones are set for astronauts including working, sleeping, sanitation, dining, healthcare and exercise.

There are three separate bedrooms and one toilet, allowing three astronauts to live for a long time, according to Zhu Guangchen, deputy chief designer of the space station system at the China Academy of Space Technology, at a press conference.

The dining zone features a retractable dining table and equipment to heat or refrigerate food, as well as supplying drinking water, Zhu said, "The exercise area is equipped with space treadmills and bikes."

He said the core module has a life support system to regenerate oxygen and dispose carbon dioxide and hazardous gas as well as recycle water. "It reduces the load of consumption goods sent to space, prolonging the stay of the astronauts."

There are also air conditioners to ensure the temperature, humidity and working temperature of equipment are within the appropriate range, Zhu added.

According to Bo, astronauts can surf the internet or call anyone on Earth with a network that runs at a maximum speed of 10 gigabits per second.

"The network that astronauts use is no difference from what we use on the ground."

C. Presentation

After you have read the article, please choose one of the following topics to develop your ideas. Make a presentation with PowerPoint to the class.

Topic 1 Designing the Astronauts' Living Space

Imagine yourself as a space station designer tasked with creating the living space for astronauts on China's space station. Your objective is to design a comfortable and functional environment that meets the needs of astronauts during their missions.

- Outline the key features and design elements of the living space, including sleeping quarters, dining area, sanitation facilities, exercise equipment, and recreational amenities.
- Explain how these features contribute to the well-being and productivity of astronauts during their time in space.

Topic 2 Sustainability and Well-being in Space Living

Explore the sustainability and well-being aspects of living in space on China's space station. Your presentation will examine the environmental systems, health considerations, and recreational opportunities that contribute to astronauts' physical and mental well-being during extended missions. You will also discuss the technological innovations and operational strategies implemented to ensure the long-term sustainability of space habitats.

Note:

- Craft a visually appealing PowerPoint with appropriate colors and images.
- Keep each slide concise, using fewer than 10 words.
- Use Pictures, illustrations or forms to make your point.
- Emphasize positive concepts and messages throughout your presentation.
- Aim to deliver your presentation smoothly, without relying on notes, within a timeframe of 4-5 minutes.

 Video

Video

A. Before You Watch

Read out the words below. Choose a word or a phrase on the left of the box and a word on the right of the box to form an appropriate expression.

1. achieve	A. staging
2. solid	B. propellant
3. strike	C. booster
4. rocket	D. devices
5. gimbaled	E. orbit
6. high precision	F. thrust

1. _____ 2. _____ 3. _____
4. _____ 5. _____ 6. _____

B. While You Watch

Exercise 1: Discussion

How many phases does rocket staging experience during a rocket launching? Which phase is the most crucial one?

Exercise 2: Dictation

Fill in the blanks with words and expressions you have heard from the video.

This unit of the rocket is called the rocket engine. The rocket engine we have discussed is specifically called the liquid propellant rocket engine. They are the most (1) _____ rocket proportion system available. The fuel and oxidizer required for the rocket engine are stored in two large tanks as shown. During liftoff, the thrust produced by the main engine (2) _____. So usually a few solid propellant strike boosters are used to assist the liftoff. You can see more details about solid propellant rockets here.

The rocket starts with zero speech at ground, but it should (3) _____ to a final speed of around 28,000 kilometers per hour to successfully (4) _____. The solid propellent strike boosters are burnt off very rapidly, so to reduce the weight of the rocket they are abandoned after the burn off. This process is known as (5)_____. When the main engine is burnt off, it is also abandoned, and the next engine takes over the charge. In this way, the rocket's weight is greatly reduced, thus greater acceleration can be achieved. Finally, after a few stages of operation, the payload is put into the (6) _____ . Rocket staging up to five has been successfully tested. You might be wondering how the rocket is able to maneuver itself to reach its (7) _____. The most (8) _____ is called gimbaled thrust. Here the rocket nozzle is (9) _____ tilted by high precision devices. It is clear that any (10) _____ from its normal angel will produce torque which will make the rocket's body turn. After achieving enough turn, the gimbal angel is set to zero.

Summary and Reflection

Now that you have completed the chapter of aerospace, it's time to reflect on your learning and ensure you have met the goals set for the chapter. Follow these steps to complete the checklist:

- Carefully read through the checklist provided, which outlines the key learning objectives and goals of the Aerospace chapter. For each item on the checklist, evaluate your own understanding and progress by checking the corresponding box.
- If you feel confident in your understanding and achievement of the goal, check the box; If you believe there are areas where you need further improvement or clarification, leave the box unchecked.

1. **Understanding of Aerospace Engineering:**
   ☐ Have I deepened my understanding of aerospace engineering concepts, including rocket staging, space station construction, and debris management strategies?
   ☐ Can I explain these concepts clearly to others, highlighting key components such as rocket propulsion systems and space station modules?
2. **Critical Thinking Skills:**
   ☐ How effectively did I apply critical thinking skills to analyze the feasibility and implications of aerospace technologies in different scenarios?
   ☐ Did I identify potential limitations and ethical considerations associated with space exploration and rocket launches?
3. **Reading Comprehension, Vocabulary, and Language Proficiency:**
   ☐ Have I improved my reading comprehension, vocabulary, and language proficiency through exercises focused on technical texts and terminology related to aerospace engineering?
   ☐ Can I accurately define and use technical terms associated with rocket propulsion, space station architecture, and debris mitigation strategies in context?

4. **Collocating Words and Phrases:**

   ☐ Can I identify and analyze the collocation of words and phrases within the context of aerospace engineering to enhance my understanding of technical terminology?

   ☐ Have I practiced using these collocations in sentences to reinforce my understanding of concepts such as rocket staging and space debris management?

5. **Understanding of Impacts:**

   ☐ How did I explore the potential impacts of aerospace engineering on various industries, including entertainment, healthcare, and advertising?

   ☐ Can I articulate the broader societal implications of space exploration and satellite technology?

6. **Presentation Skills:**

   ☐ How did I develop my presentation skills by creating and delivering presentations on topics related to aerospace engineering, using tools like PowerPoint?

   ☐ Did I effectively organize information, use visual aids, and engage the audience to convey complex aerospace concepts?

7. **Reflection and Critical Analysis:**

   ☐ How did I reflect on and critically analyze the knowledge acquired throughout the chapter, particularly in relation to the ethical considerations and future applications of aerospace engineering?

   ☐ Did I consider different perspectives and evaluate the significance of aerospace advancements in various fields?

Chapter 7
Cloud Computing

> **Objectives**
> In this chapter, you should be able to:
> - Analyze and discuss the convenience cloud computing brings to our lives also problems brought in different aspects.
> - Develop critical thinking, reading, and writing skills by analyzing the problems brought by cloud computing.
> - Demonstrate the ability to summarize and categorize information from complex texts related to cloud computing using techniques such as multiple choice, mind mapping, and matching exercises.
> - Expand vocabulary and language proficiency in an academic context through exercises like glossaries, word choice, and translation tasks related to cloud computing technology and its terminology.
> - Engage in collaborative discussions and presentations to share insights and perspectives on cloud computing, its development and problems, both before and after reading relevant material.

 Before You Read

A. Discussion
Look at the pictures below and discuss with a partner.
1. What smart home devices are shown in the pictures?
2. How are these devices connected to the cloud?
3. What role do you think cloud computing plays in the smart home?
4. What would a smart home look like without cloud computing?

B. Skimming and Scanning

Browse the text and answer the questions below.

1. Which company has started investing in solar energy to mine Bitcoin?

A. Apple

B. Google

C. Facebook

D. Square

2. According to the article, what percentage of global electricity is currently used for cloud computing?

A. 1%

B. 8%

C. 25%

D. 50%

3. What is NOT true about cloud computing according to this passage?

A. Moore's Law keep the power budget in check as we scaled up our computing resources at earlier time.

B. The most immediate solution is to process more data at the edge, before it goes into the cloud.

C. There should be no worry about the situation where computing and power consumption will be strongly coupled as 60 years ago.

D. Software and hardware engineering will have to consider their design around power efficiency.

Text

Cloud Computing's Coming Energy Crisis
The Cloud's Electricity Needs Are Growing Unsustainably

1. How much of our computing now happens in the cloud? A lot. Providers of public cloud services alone take in more than a quarter of a trillion dollars a year. That's why Amazon, Google, and Microsoft maintain massive data centers all around the world. Apple and Facebook, too, run similar facilities, all stuffed with high-core-count CPUs, sporting terabytes of RAM and petabytes of storage.

2. These machines do the heavy lifting to support what's been called "surveillance capitalism": the endless tracking, user profiling, and algorithmic targeting used to distribute advertising. All that computing rakes in a lot of dollars, of course, but it also consumes a lot of watts: Bloomberg recently estimated that about 1 percent of the world's electricity goes to cloud computing.

3. That figure is poised to grow exponentially over the next decade. Bloomberg reckons that, globally, we might exit the 2020s needing as much as 8 percent of all electricity to power the future cloud. That might seem like a massive jump, but it's probably a conservative estimate. After all, by 2030, with hundreds of millions of augmented-reality spectacles streaming real-time video into the cloud, and with the widespread adoption of smart digital currencies seamlessly blending money with code, the cloud will provide the foundation for nearly every financial transaction and user interaction with data.

4. How much energy can we dedicate to all this computing? In an earlier time, we could have relied on Moore's Law to keep the power budget in check as we scaled up our computing resources. But now, as we wring out the last bits of efficiency from the final few process nodes before we reach atomic-scale devices, those improvements will hit physical limits. It won't be long until computing and power consumption will once again be strongly coupled—as they were 60 years ago, before integrated CPUs changed the game.

5. We seem to be hurtling toward a brick wall, as the rising demand for computing collides with decreasing efficiencies. We can't devote the whole of the planet's electricity generation to support the cloud. Something will have to give.

6. The most immediate solutions will involve processing more data at the edge, before it goes into the cloud. But that only shifts the burden, buying time for rethinking how to manage our computing in the face of limited power resources.

7. Software and hardware engineering will no doubt reorient their design practices around power efficiency. More code will find its way into custom silicon. And that code will find more reasons to run infrequently, asynchronously, and as minimally as possible. All of that will help, but as software progressively eats more of the world—to borrow a now-famous metaphor—we will confront this challenge in ever-wider realms.

8. We can already spy one face of this future in the nearly demonic coupling of energy consumption and private profit that provides the proof-of-work mechanism for cryptocurrencies like Bitcoin. Companies like Square have announced investments in solar energy for Bitcoin mining, hoping to deflect some of the bad press associated with this activity. But more than public relations is at stake.

9. Bitcoin asks us right now to pit the profit motive against the health of the planet. More and more computing activities will do the same in the future. Let's hope we never get to a point where the fate of the Earth hinges on the fate of the transistor.

Reading Comprehension

A. Multiple Choice
Choose the best answer for each question.

1. What does the author predict about the electricity needed for cloud computing in the 2020s?

   A. It will remain stable.

   B. It will decrease.

C. It will increase exponentially.

D. It will double.

2. What challenge does the author mention regarding Moore's Law?

A. It is no longer applicable to modern computing.

B. It cannot keep up with the rising demand for computing.

C. It will lead to a decrease in computing efficiency.

D. It cannot be relied on to manage power consumption.

3. What does the author suggest as an immediate solution to the rising demand for computing?

A. Processing more data at the edge.

B. Building more data centers.

C. Developing more efficient CPUs.

D. Limiting the use of cloud computing.

4. What does the author mean by "the nearly demonic coupling of energy consumption and private profit" in paragraph 8?

A. Cryptocurrencies are very profitable.

B. Cryptocurrencies consume a lot of energy.

C. Cryptocurrencies are environmentally friendly.

D. Cryptocurrencies are difficult to mine.

5. What is the main concern of the author regarding the future of computing?

A. Its impact on the environment.

B. Its ability to meet rising demand.

C. Its potential to create new jobs.

D. Its efficiency in data processing.

B. Mind Map

How many main parts do you think the article is composed of? Fill in the blanks with the details you read from the article.

Chapter 7 Cloud Computing

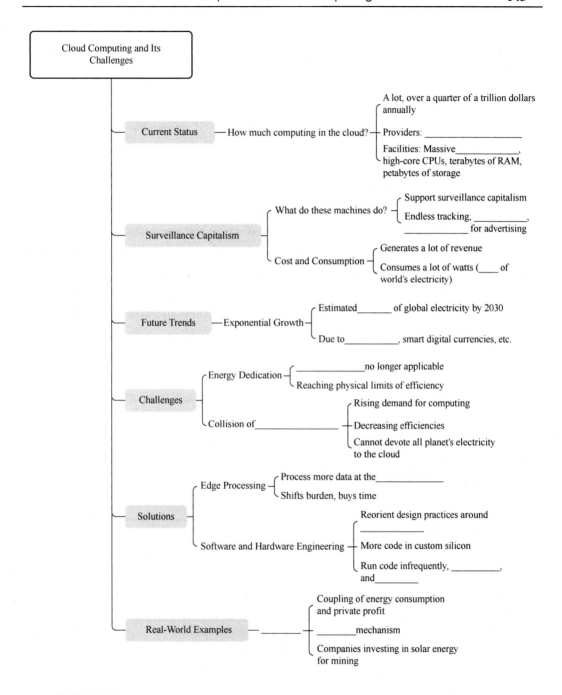

C. Matching

Read the text and decide which paragraph mentions the following information? Write the number of the paragraph before each sentence.

_____ 1. Software and hardware engineering will have to reorient their design practices towards power efficiency to address the energy challenges.

_____ 2. The electricity consumption for cloud computing is expected to grow exponentially in the next decade, potentially reaching 8% of global electricity by 2030.

_____ 3. The increasing demand for computing and decreasing efficiency pose a challenge, as we cannot devote all of the planet's electricity to support cloud computing.

_____ 4. A significant amount of computing is now done in the cloud, generating significant revenue for providers like Amazon, Google, and Microsoft.

_____ 5. Immediate solutions involve processing more data at the edge before sending it to the cloud, but this only buys time for rethinking management strategies.

_____ 6. More computing activities in the future may pit the profit motive against the health of the planet, presenting a challenge for sustainable development.

_____ 7. These cloud computing machines support "surveillance capitalism" while consuming a significant amount of electricity, estimated to be around 1% of the world's total.

_____ 8. The energy consumption and private profit coupling in cryptocurrencies like Bitcoin highlights a possible future where computing activities have significant environmental impacts.

_____ 9. As we reach the physical limits of improving computing efficiency, the coupling between computing and power consumption is likely to increase.

D. Cloze

The information below is a summary of the text. Complete the summary by filling in the blanks with the words and phrases provided.

A. computational efficiency	F. data centers
B. cryptocurrencies	G. private profit
C. custom silicon	H. edge computing
D. revenue	I. algorithmic targeting
E. energy consumption	J. diminishing efficiency

Cloud computing has become pervasive, generating significant (1)_____ for providers like Amazon, Google, and Microsoft, who maintain vast (2)_____ worldwide. These centers power "surveillance capitalism" through data tracking and (3)_____, but they also consume vast amounts of electricity—currently estimated at 1% of global power usage.

The demand for cloud computing is expected to grow exponentially, potentially requiring up to 8% of global electricity by 2030. However, as we reach the physical limits of (4)_____, the decoupling of computing power and (5)_____ is ending. This impending collision between rising demand and (6)_____ poses a significant challenge.

Immediate solutions involve (7)_____ to reduce cloud processing, but this only delays the need for more sustainable solutions. Software and hardware design will need to shift to prioritize power efficiency, with more code optimized for (8)_____ and minimal, asynchronous execution.

(9)_____ like Bitcoin already highlight the problematic coupling of energy consumption and (10)_____. As computing activities expand, we must ensure that the future of the planet does not hinge on the fate of the transistor.

 Language Building

A. Glossary

Proper Nouns
Amazon 亚马逊，是美国的网络电子商务公司，总部位于华盛顿州的西雅图，是网络上较早开始经营电子商务的公司之一，成立于 1994 年。一开始，该公司只经营书籍销售业务，现在则涉及范围相当广的其他产品。
Google 谷歌，成立于 1998 年 9 月，由拉里·佩奇和谢尔盖·布林共同创建，业务包括互联网搜索、云计算、广告技术等，同时开发并提供大量基于互联网的产品与服务。
Microsoft 微软，是一家美国跨国科技企业，由比尔·盖茨和保罗·艾伦于 1975 年 4 月创立。公司总部设立在华盛顿州雷德蒙德，主要业务为研发、制造、授权和提供广泛的计算机软件服务。
Apple 苹果，由史蒂夫·乔布斯、斯蒂夫·盖瑞·沃兹尼亚克和罗纳德·杰拉尔德·韦恩等人于 1976 年 4 月创立，总部位于加利福尼亚州库比蒂诺。

Facebook

脸书，创立于 2004 年 2 月 4 日，总部位于美国加利福尼亚州门洛帕克，2021 年 10 月，Facebook 正式更名为 Meta。

CPU（Central Processing Chapter）

中央处理器，是计算机系统的运算和控制核心，是信息处理、程序运行的最终执行单元。CPU 自产生以来，在逻辑结构、运行效率及功能外延上取得了巨大发展。

RAM（Random Access Memory）

随机存取存储器，也叫主存，是与 CPU 直接交换数据的内部存储器。它可以随时读写（刷新时除外），而且速度很快，通常作为操作系统或其他正在运行中的程序的临时数据存储介质。

Bloomberg

彭博，是全球商业、金融信息和财经资讯的领先提供商，由迈克尔·彭博于 1981 年创立，总部位于美国纽约市曼哈顿。

Moore's Law

摩尔定律，是英特尔创始人之一戈登·摩尔的经验定律，其核心内容为：集成电路上可以容纳的晶体管数目每经过 18 个月到 24 个月便会增加一倍。换言之，处理器的性能大约每两年翻一倍，同时价格下降为之前的一半。

摩尔定律并非自然科学定律，而是一种经验总结，它在一定程度揭示了信息技术进步的速度。

Bitcoin

比特币，最初由中本聪在 2008 年 11 月 1 日提出，并于 2009 年 1 月 3 日正式诞生。比特币是一种 P2P 形式的数字货币，其交易记录公开透明。

Square

Square 是美国一家移动支付公司，其创始人是杰克·多尔西。Square 用户（消费者或商家）利用 Square 提供的移动读卡器，配合智能手机，可以在任何移动或无线网络状态下，通过应用程序匹配刷卡消费，大大降低了刷卡消费支付的技术门槛和硬件需求。

Academic Words	
massive (a.)	大量的，大规模的

terabyte (n.)	（计算机）万亿字节，兆兆字节
petabyte (n.)	拍字节，千万亿字节
surveillance (n.)	监视，监察
capitalism (n.)	资本主义
profiling (n.)	资料收集，剖析研究
algorithmic (a.)	［数］算法的
rake in	迅速大量取得；大量地搜刮（钱财）
consume (v.)	消耗
poised (a.)	做好准备的
exponentially (adv.)	以指数方式
reckon (v.)	估计，估算
conservative (a.)	保守的
spectacles (n.)	眼镜
seamlessly (adv.)	无缝地
blending (a.)	混合的
transaction (n.)	交易，买卖，业务
dedicate (v.)	致力于，献身于，献给
in check	受控制的；受抑制的
scale up	按比例放大；按比例增加
wring out	绞出，扭干，榨取
consumption (n.)	消耗
hurtle (v.)	猛冲，猛烈碰撞
collide (v.)	冲突
reorient (v.)	使适应；再调整
silicon (n.)	硅
asynchronously (adv.)	异步地；不同时地
metaphor (n.)	隐喻，暗喻
confront (v.)	面对，面临
realm (n.)	领域，范围
demonic (a.)	恶魔的
Mechanism (n.)	机制
cryptocurrency (n.)	加密电子货币
deflect (v.)	使转向
at stake	处于危险中；在紧要关头
hinge on	取决于；以……为转移

B. Words and Phrases

Exercises 1 Word Choice

Use the words or phrases in the box to finish the sentences.

| massive | rake in | exponentially | scale up | wring out |
| hurtle | asynchronously | reorient | spy | hinge on |

1. The quantity of chemical pollutants has increased _____.
2. It can be used synchronously or _____.
3. We need a work vacation or two every year in order to _____ ourselves and reestablish our sense of perspective.
4. The privatization allowed companies to _____ huge profits.
5. Going through his list of customers is a _____ job.

Exercise 2 Phrases

Match the words provided below with appropriate one in the box.

1. software _____
2. cloud _____
3. augmented _____
4. atomic _____
5. power _____

| reality |
| scale |
| engineering |
| consumption |
| computing |

Exercise 3 Sentence Completion

Complete the sentences by filling in the blanks with phrases in the above exercise.

1. Indeed, on the _____, perpetual motion is a reality.
2. Can _____ help save the print publishing industry?
3. _____ is not computer science — it is bigger.
4. Every home cuts _____ by 50%, mostly by using low-power appliances and solar panels.
5. Server virtualization created _____.

Exercise 4 Translation

Translate the sentences by using the words and phrases you have learned in the above two exercises.

1. 在今天的软件工程实践中，模式是相当重要的一部分。
___

2. 爆炸在地面留下了一个巨大的坑。
___

3. 这个特性使它可以实时地收集电力消耗数据。
___

4. 报告发现地球轨道上的碎片成几何数增长。
___

5. 如果没有事件排序，这些消息将会被异步处理。
___

C. Collocation

Exercise 1 Modifiers
Match the English phrase with the correct Chinese definition.

chromatic scale	A. 评定量表
atomic scale	B. 原子标度
grand scale	C. 前所未有的规模
global scale	D. 时间标尺
rating scale	E. 大规模
unprecedented scale	F. 半音音阶
time scale	G. 全球范围

Exercise 2 Blank Filling
Scan the text and complete the sentences containing the word "scale".

1. But now, as we wring out the last bits of efficiency from the final few process nodes before we reach _____ devices ...

2. Indeed, on a _____, fertilizer manufacturing consumes about 3%-5% of the world's annual natural gas supply.

3. _____ is generated by synthesizing a number of precision clocks.

4. New Yorkers built their city on a _____.

5. The abstinence symptom _____ was used as marking and grading of symptoms.

6. Play a _____ for the full range of the instrument in long tones, taking the time to check and correct each note's intonation.

7. Other parts of the globe, a variety of natural disasters broke out on an _____.

Exercise 3   Translation

Translate the sentences below from Chinese to English using "scale" and its collocation in this section.

1. 这个领域确实是正在以前所未有的规模发展的新领域。
_____

2. 它将在全球范围内推动技术进步。
_____

3. 在一张单子上列出所有属性，利用五分评价尺度来对绩效进行评价。
_____

4. 许多科学家为地质时间尺度的建立提供了有价值的研究和投入。
_____

5. 我们应该大规模普及科学。
_____

D. Terminology

Exercise 1 Table Filling

Read the article and find the English technical terms according to the Chinese equivalents.

English Technical Terms	Chinese Equivalents
	云计算
	增强现实眼镜
	软件工程
	电力消耗
	数字货币
	工作量证明机制

Exercises 2   Blank Filling

Use the terms in the above exercise to complete the sentences below.

1. Bloomberg recently estimated that about 1 percent of the world's electricity goes to_____.

2. After all, by 2030, with hundreds of millions of _____ streaming real-time video into the cloud ...

3. ... and with the widespread adoption of smart _____ seamlessly blending money with code ...

4. It won't be long until computing and _____ will once again be strongly coupled—as they were 60 years ago, before integrated CPUs changed the game.

5. _____ will no doubt reorient their design practices around power efficiency.

6. We can already spy one face of this future in the nearly demonic coupling of energy consumption and private profit that provides the _____ for cryptocurrencies like Bitcoin.

Critical Reading and Writing

A. Brainstorming

Work in groups. Fill in the table according to the instruction.

Energy consumption is one of the major issues raised by the development of cloud computing mentioned in the article. Please list three other issues besides energy consumption and summarize your reasons, then put each in one sentence.

Issue 1: _____
Issue 2: _____
Issue 3: _____

B. Critical Reading

Exercise 1: The following are the answers to the question provided by ERNIE Bot. Read the introduction, and then match the headings and the corresponding content in the body part.

Heading A: Reliability and availability

Heading B: Cost issues

Heading C: Compatibility and integration issues

Heading D: Management and operation challenges

Heading E: Data privacy and compliance

Heading F: Security issues

Heading G: Performance issues

The rapid development and widespread adoption of cloud computing has indeed brought many conveniences, but it has also exposed some issues and challenges. Here are some common problems in the use and development of cloud computing:

a. _____:

As data is stored in the cloud, cloud service providers bear more security responsibilities. However, security threats such as hacking attacks, data breaches, and malware always exist, posing potential risks to users and businesses.

b. _____:

In cloud computing, data may be stored on servers located in multiple different countries and regions, involving different laws, regulations, and privacy policies. Therefore, cloud service providers need to comply with laws and regulations in multiple countries and regions to ensure data compliance and privacy protection.

c. _____:

Due to the complexity and uncertainty of the cloud computing environment, service interruptions or failures may occur occasionally. And these can cause losses to users and businesses.

d. _____:

The performance of cloud computing can be affected by various factors, such as network latency, bandwidth limitations, and resource contention. These factors may slow down the response speed of applications, affecting user experience. Additionally, if cloud service providers do not plan resource allocation properly, it may lead to performance bottlenecks.

e. _____:

While cloud computing can reduce IT costs for businesses, it may also bring some additional costs. First, businesses need to pay cloud service providers for services, which may include infrastructure fees, software license fees, data storage fees, etc. Second, if businesses do not plan their cloud computing resource usage properly, it may lead to resource waste or insufficient resources, thus increasing costs.

f. _____:

Different cloud service providers may adopt different technologies, standards, and protocols, which may lead to some issues for businesses. Additionally, if businesses need to integrate cloud computing with other systems or applications, they may face some technical challenges.

g. _____:

In a cloud environment, whether private, public or hybrid, it takes a group of individuals or teams with the expertise and experience to ensure proper operation and performance optimization. However, managing and operating cloud computing environments requires certain technical capabilities and experience, which may pose some challenges for businesses.

> To address these issues, cloud service providers and businesses need to take a series of measures, such as strengthening security management, improving data privacy protection, optimizing resource allocation, improving performance and reliability, reducing costs, enhancing compatibility and integration capabilities, and improving management and operation capabilities. At the same time, it is also necessary to strengthen international cooperation and standardization efforts to promote the healthy development of cloud computing technology.

Exercise 2: Do you agree with ERNIE Bot on the above ideas? Can you think of solutions to these problems? Share your ideas with the class.

C. Essay Writing

Having engaged in discussions and vocabulary preparation, you've likely generated numerous insightful ideas. Now, it's time to reflect on these ideas and the knowledge you've acquired by crafting an essay. Use the following instructions to guide your writing process:

 Topic

Problems brought by cloud computing in … (a specific aspect)

Background Information

In the digital wave sweeping the world today, cloud computing has become an indispensable technical support for enterprises, organizations and even individuals. With its powerful computing power, flexible resource allocation and efficient cost effectiveness, it has reshaped our understanding and application of information technology. With the wide application of cloud computing technology, more and more enterprises are beginning to migrate core services to the cloud, expecting to achieve flexible resource expansion, fine cost control and fast service delivery through cloud computing.

However, just like any technology has two sides, cloud computing brings endless convenience and efficiency gains, but also comes with a series of complex and thorny issues. What we have to face is that with the widespread application of cloud computing, its complexity and uncertainty are also increasing. From data security to service reliability, from cost-effectiveness to technological innovation, every aspect requires in-depth consideration and discussion.

Instructions

In an essay of approximately 300-450 words, analyze how cloud computing brings problems in a particular aspect. Provide specific evidence and arguments to support your opinion.

Your essay should include the following components:

Introduction (approximately 50-75 words): Briefly introduce the topic and provide context for your analysis.

- State your thesis or main argument regarding what problems cloud computing will bring problems to in a certain aspect.

Body Paragraphs (approximately 250-350 words): Present your argument in detail, supported by specific evidence and examples in 3 separate paragraphs.

- Describe in detail the problems in the selected areas, such as data security, privacy protection, service reliability, cost effectiveness, etc.
- Analyze the impact of these issues on individuals, enterprises and even society as a whole.
- Use specific cases to support your point of view and make your argument more convincing.
- Explore the causes of problems, such as technical defects, improper management, lack of policies, etc.

Conclusion (approximately 50-75 words): Summarize your main points and restate your thesis in light of the evidence presented.

- Restate the importance of the chosen issue.
- Look into the future development trend and prospect of cloud computing.
- Propose a solution or suggestion to the selected problem.

* Ensure that your essay is well-structured, logically organized, and supported by evidence from reputable sources. Use clear and concise language, and proofread your work carefully for grammar, punctuation, and spelling errors.

Chapter 7　Cloud Computing　　• 155 •

 Discussion and Presentation

A. Group Discussion
Exercise 1: Think and discuss the questions below.
1. What's is the development trend of China's cloud computing market and which industries in China have widely adopted cloud computing?
2. What competitive advantages and disadvantages do Chinese cloud computing enterprises have in the international market? How should China further enhance the international competitiveness of these enterprises?
3. How does cloud computing promote the development of China's digital economy?
4. How does cloud computing help Chinese Small and Medium-sized Enterprises (SMEs) reduce costs and improve efficiency?
5. How does cloud computing combine with Artificial Intelligence (AI) technology to bring new opportunities for China's development?

Exercise 2: Read the article below. Search for any information related to the above questions. Do you think the answers provided in the article match yours? Share your ideas with the class.

B. Extended Reading-Cloud Computing in China
Read the article, highlight any ideas or language that will help you in your presentation in the next session.

China Embraces Another Breakthrough in Cloud Computing

Alibaba Cloud recently unveiled the Cloud Infrastructure Processing Chapter (CIPU) designed to power cloud-native data centers. The CIPU is expected to replace the CPU, or the central processing chapter, as the core of a new generation of cloud computer system.

Over the past decades, the U.S. has constantly held a leading position globally in the information industry. For instance, the Electronic Numerical Integrator and Computer, better known as ENIAC, developed by the University of Pennsylvania became the world's first programmable general-purpose electronic computer on Feb. 14, 1946, raising the curtain of the information age. Sixty years later, Amazon's Simple Storage Service, or Amazon S3 ushered in the era of cloud computing as the world's first cloud computing service. "The

CIPU has completely changed the structure of the last-gen computing architecture, making China a world leader in basic technology," said Zheng Weimin, academician of the Chinese Academy of Engineering and head of the Department of Computer Science and Technology, Tsinghua University.

The CIPU and CPU are completely different despite the minor difference in their acronyms. The CPU, a super large integrated circuit, is the core of a computer's computing and controlling operations. Its major function is explaining orders and processing software data. However, Alibaba Cloud's CIPU is the hub of control and acceleration for cloud computing.

In the digital era, super large computing scenarios and power are especially important for populous countries like China, and the demand for cloud computing is constantly rising as the internet of things approaches. To meet the ever-increasing computing demand, China officially kicked off a project that "processes eastern data in the west" in February, 2022, trying to build a new type of computing network that combines data centers, cloud computing, and big data through 10 national data center clusters.

Data center construction calls for huge and systematic efforts. As a senior player in the cloud computing sector, Alibaba Cloud finds that data centers with traditional CPU at the core are largely hindered by wastage of processing power. It is because as the data centers grow bigger, the EAST-WEST traffic gets bigger, too, which makes cloud computing turn from a business processing center to a data processing center, and the CPU was more engaged in the network business, rather than computing. It means many servers deployed at data centers are going to waste.

Therefore, Alibaba Cloud developed the CIPU, a processor for new cloud data centers. It transfers the computing, storage and internet resources of data centers to the cloud and enables hardware acceleration, and is also connected to Alibaba's Apsara system, turning millions of servers around the world into a super computer.

It is proved that there's a huge increase in the computing power with the CIPU and Apsara at the core. In mainstream computing scenarios, the Nginx performance has been raised by 89 percent, the Redis performance 68 percent and the MySQL performance 60 percent. In big data and AI scenarios, there's a 30 percent rise in both AI learning and the Spark engine. The consumption of virtualization is down by 50 percent, while the startup speed of virtualization containers has increased by three-and-a-half times.

The improvement is not limited to performance. Under traditional IT frameworks, the performance and manufacturing difficulty of silicon-based chips have almost reached the limit, so it's quite uncertain when the next breakthrough will be made. Therefore, more and more researchers resorted to cloud computing, hoping to bring about a huge increase in computing power through a new type of computing structure, rather than the silicon-based chips alone.

With the CIPU, Alibaba Cloud is forging ahead neck and neck with other international IT giants, according to Zheng, who added that the processing chapter has gained an upper hand for China in the definition of cloud computing and changed the rules set by the West. He said he believes China will gain a foothold in the next era of technology.

C. Presentation

After you have read the article, please choose one of the following topics to develop your ideas. Make a presentation with PowerPoint to the class.

Topic 1　Data center and resource optimization in cloud computing:
- Discuss the role of data centers in cloud computing and why they are critical for processing large amounts of data and delivering efficient services.
- Analyze the challenges faced by traditional data centers.
- Explains how CIPU enables hardware acceleration and resource optimization in the data center.
- Explore the possible shape and challenges of the future data center as new technologies continue to emerge.

Topic 2　China's role in the global cloud computing race
- Analyze China's position and role in the global cloud computing competition.
- Analyze the contribution of Alibaba and other Chinese enterprises in the field of cloud computing.
- Discuss how China can achieve greater influence in the global cloud computing sector through technological innovation and international cooperation.
- Discuss the challenges and opportunities that China may face in the future development of cloud computing.

Note:
- Craft a visually appealing PowerPoint with appropriate colors and images.
- Keep each slide concise, using fewer than 10 words.

- Use pictures, illustrations or forms to make your point.
- Emphasize positive concepts and messages throughout your presentation.
- Aim to deliver your presentation smoothly, without relying on notes, within a timeframe of 4-5 minutes.

 Video

A. Before You Watch

Read out the words below. Choose a word in the box to form on appropriate expression.

| Learning | big-data | infrastructure | capital | data | disaster | center |

- _____ service
- _____ backup
- _____ recovery
- data _____
- machine _____
- _____ expense
- _____ analytics

B. While You Watch

Exercise 1: Discussion

Can you think any convenience cloud computing has brought to our lives?

Exercise 2: Dictation

Fill in the blanks with words and expressions you have heard from the video.

Cloud computing is the on-demand delivery of IT resources via the Internet, with pay as you go pricing. Instead of buying, owning and maintaining physical (1) _____, you can access technology services such as computing, power, storage and databases on an as needed basis from a cloud provider like Amazon Web services. Organizations of every type, size and industry are using the cloud for a wide variety of use cases, such as (2) _____, disaster recovery, email, (3) _____, software development and testing, (4) _____ and customer facing web applications.

For example, healthcare companies are using the cloud to develop more (5) _____ for patients. Financial services companies are using the cloud to power real-time (6) _____. And video game makers are using the cloud to deliver online games to millions of players around the world.

With cloud computing, your business can become more agile, reduce costs, instantly (7) _____ globally in minutes. Cloud computing gives you instant access to a broad range of technologies, so you can innovate faster and build nearly anything you can imagine, from (8) _____ such as compute storage and databases to (9) _____, machine learning, data analytics and much more. You can deploy technology services in a matter of minutes and get from idea to implementation several orders of magnitude faster than before. This gives you the freedom to experiment and test new ideas to differentiate customer experiences and transform your business, such as adding (10) _____ to your applications, in order to personalize experiences for your customers and improve their engagement.

You don't need to make large (11) _____ in hardware and overpaid for capacity you don't use. Instead, you can trade capital expense for variable expense and only pay for IT as you consume it. With cloud computing, you access resources from the cloud in real time as they're needed. You can scale these resources up and down to grow or shrink capacity instantly as your business needs change. Cloud computing also makes it easy to expand to new regions and deploy globally in minutes. For example, Amazon Web services has infrastructure all over the world, so you're able to deploy your application in multiple physical locations in just a few clicks. Putting applications in closer proximity to end users (12) _____ and improves their experience, no matter your own location, size or industry. The cloud frees you from managing infrastructure and data centers, so you can focus on what matters most to your business.

Summary and Reflection

Now that you have completed the chapter of cloud computing, it's time to reflect on your learning and ensure you have met the goals set for the chapter. Follow these steps to complete the checklist:

- Carefully read through the checklist provided above, which outlines the key learning objectives and goals of the cloud computing chapter. For each item on the checklist, evaluate your own understanding and progress by checking the corresponding box.

- If you feel confident in your understanding and achievement of the goal, check the box; If you believe there are areas where you need further improvement or clarification, leave the box unchecked.

1. **Understanding of Cloud Computing:**
   ☐ Have I gained a deeper understanding of cloud computing technology, including its basic concepts, the convenience it brings, and the problems caused by it?
   ☐ Can I clearly explain these to others?

2. **Critical Thinking Skills:**
   ☐ How effectively did I apply critical thinking skills to tasks such as skimming, scanning, and matching to extract key information from technical texts?
   ☐ Did I identify relevant information and draw logical conclusions?

3. **Reading Comprehension, Vocabulary, and Language Proficiency:**
   ☐ Have I improved my reading comprehension of cloud-computing-related texts, expanding my vocabulary to include terms?
   ☐ Can I accurately define and use these technical terms in appropriate contexts?

4. **Collocating Words and Phrases:**
   ☐ Can I identify and analyze the collocation of cloud-computing-specific words and phrases to enhance my comprehension of the subject matter?
   ☐ Have I practiced using these collocations in sentences to convey information about cloud computing?

5. **Understanding of Energy Consumption Issues:**
   ☐ How did I approach the issue of energy consumption in cloud computing? Did I consider solutions like data center efficiency and green computing?
   ☐ Can I articulate the severity and potential impacts of energy consumption issues in cloud computing?

6. **Understanding of Cloud Computing Development in China:**
   ☐ What is my understanding of the current status and trends of cloud computing development in China? Did I read and analyze relevant articles?
   ☐ Can I present my own views and insights on the innovations and developments in cloud computing in China?

7. **Presentation Skills:**
   ☐ How did I develop my presentation skills by creating and delivering presentations on topics related to cloud computing, using tools like PowerPoint?
   ☐ Did I effectively organize information, use visual aids, and engage the audience through interactive elements?
8. **Reflection and Critical Analysis:**
   ☐ How can I reflect on and critically analyze the knowledge gained throughout the chapter, especially regarding its applications and impacts?

Chapter 8
Smart Farm

> **Objective**
> In this chapter, you should be able to:
> - Understand the concept of "smart agriculture" and its use of digital technology.
> - Analyze the security issues specific to smart agriculture, such as sensor failures and control system intrusions.
> - Discuss the research challenges associated with developing security solutions for smart agriculture, including the need for new datasets and the emergence of new security issues.
> - Summarize the main points and arguments presented in the cybersecurity report.
> - Develop critical thinking skills by evaluating the effectiveness of existing security and privacy countermeasures for smart agriculture.
> - Expand vocabulary related to cybersecurity and agriculture.
> - Engage in discussions about the implications of cybersecurity vulnerabilities in smart agriculture.
> - Develop presentations on the topic of cybersecurity in smart agriculture, focusing on potential solutions and research gaps.

 Before You Read

A. Discussion

Look at the pictures below and discuss with a partner.
1. What scenes do the two pictures describe?
2. What is unusual about the farms in the pictures?
3. What technology are required to enable the scenarios described in the pictures?

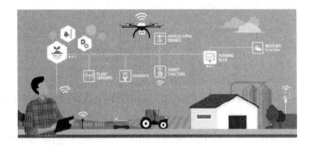

B. Skimming and Scanning

1. What comes with the digital technology in the era of "smart agriculture"?

A. IoT

B. Big data

C. Network security loopholes

D. Cloud Computing

2. Which of the followings is NOT the goal of Agriculture 4.0?

A. Increased efficiency

B. Powerful data analysis

C. More intelligent automation

D. Greenhouse cultivation

3. What technologies are available to alleviate some of the existing problems?

A. Blockchain

B. Edge computing

C. Artificial Intelligence

D. Network intrusion detectors

Text

Cybersecurity Report: "Smart Farms" Are Hackable Farms

1. Some have dubbed this the era of "smart agriculture"—with farms around the world scaling up their use of the Internet, IoT, big data, cloud computing and artificial intelligence to increase yields and sustainability. Yet with so much digital technology, naturally, also comes a heightened potential cybersecurity vulnerabilities.

2. There's no scaling back smart agriculture either. By the end of this decade we will need the extra food it produces—with world's population projected to cross 8.5 billion, and more than 840 million people are affected by acute hunger. Unless smart agriculture can dramatically increase the global food system's efficiency, the prospect of reducing global malnutrition and hunger—let alone the ambitious goal of zero hunger by 2030—appears very difficult indeed.

3. Agriculture 4.0 aims not just at growing more food but also at increased efficiency, more powerful data analysis, and more intelligent automation and decision-making.

4. Although smart agriculture has been extensively studied, "the security issues around smart agriculture have not," says Xing Yang from Nanjing Agricultural University in China. Research in the field to date, he adds, has mostly involved applying conventional cybersecurity wisdom to agricultural technology. Agricultural cybersecurity, by contrast, he says, is not given enough attention.

5. Yang and his colleagues surveyed the different kinds of smart agriculture, as well as the key technologies and applications specific to them. Agricultural IoT applications have unique characteristics that give rise to security issues, which the authors enumerate and suggest countermeasures for. (Their research was published in a recent issue of the journal IEEE/CAA Journal of Automatica Sinica.)

6. For example, while field agriculture might be subject to threats from damage to the facility, poultry and livestock breeding may face sensor failures, and greenhouse cultivation could

face control system intrusions. All of these could result in damage to the IoT architecture, both hardware and software, leading to failure or malfunction in farming operations. Plus, there are threats to data acquisition technologies—malicious attacks, unauthorized access, privacy leaks, and so on—while blockchain technologies can be vulnerable to access control failure and unsafe consensus agreement.

7. In Yang's opinion, the most pressing security problems in smart agriculture involve the physical environment, such as plant factory control system intrusion and Unmanned Aerial Vehicle (UAV) false positioning. "The network for rural areas is not as good as that of cities," he says, "which means that the network signals in some areas are poor, which leads to…false base station signals."

8. The researchers also paid extra attention to agricultural equipment as potential security threats, something that recent studies have not done. "Considering that the deployment of IoT devices in farmland is relatively sparse and cannot be effectively supervised, how to ensure the physical security of these devices is a challenge," Yang says. "In addition, the delay caused by long-distance signal transmission also increases the risk of Sybil attacks which is transmitting malicious data through virtual nodes."

9. In their experiments with solar insecticidal lamps, for instance, they found that the lamp's high voltage pulse affects the data transmission from Zigbee-based IoT devices and data acquisition sensors. Thus, Yang says, to minimize unnecessary losses, it's important to study each device in the context of how it's actually deployed in the field, including the possible safety risks of specific agricultural equipment.

10. The study also summarizes existing security and privacy countermeasures suitable for smart agriculture, including authentication and access control protocols, privacy-preserving frameworks, robust intrusion detection systems, and cryptography and key management. Yang is optimistic that the application of existing technologies—such as edge computing, artificial intelligence and blockchain—can be used to mitigate some of the existing problems. He says that AI algorithms can be developed that might detect the presence of malicious users, while existing industrial security standards can be applied to design a targeted security scheme for agricultural IoT.

11. This represents a significant research challenge, he says, because current datasets used in deep-learning approaches are not based on smart agriculture environments. Therefore, new datasets are required to build network intrusion detectors in a smart agriculture environment. "These new technologies can help the development of smart agriculture and solve some of the existing security problems," Yang says, "but they have loopholes, so they also bring new security issues."

Reading Comprehension

A. Multiple Choice

Choose the best answer for each question.

1. What term is used to describe the integration of Internet, IoT, big data, cloud computing, and artificial intelligence in agriculture?

   A. Agriculture 4.0
   B. Smart Farm
   C. Digital Agriculture
   D. Precision Agriculture

2. According to the article, why is the security of smart agriculture systems often overlooked?

   A. Lack of interest from researchers
   B. Insufficient funding for cybersecurity research
   C. Focus on applying conventional cybersecurity wisdom
   D. Limited awareness of potential security vulnerabilities

3. What does the article identify as a potential security threat in the context of agricultural IoT applications?

   A. Unauthorized access to data acquisition technologies
   B. Failure of blockchain technologies
   C. Malicious attacks on urban networks
   D. Lack of network signals in rural areas

4. What aspect of agricultural equipment does the article highlight as a potential security

concern?

    A. Deployment density of IoT devices

    B. Limited supervisory capabilities

    C. High voltage pulse from insecticidal lamps

    D. Long-distance signal transmission delays

5. According to the article, what technologies are proposed as potential solutions to mitigate security problems in smart agriculture?

    A. Edge computing, artificial intelligence, and blockchain

    B. Cloud computing, IoT, and big data analytics

    C. Augmented reality, virtual reality, and machine learning

    D. Robotics, automation, and sensor networks

B. Mind Map

How many main parts do you think the article is composed of? Group the paragraphs and fill in the blanks with the information you read from the article.

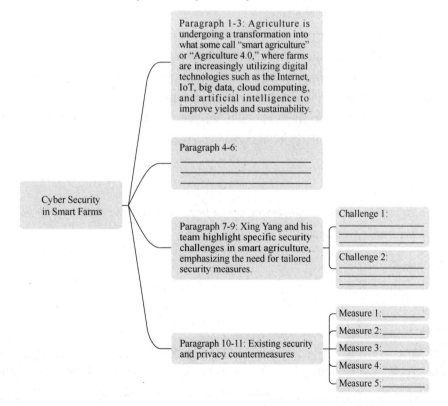

C. Matching

Read the text and decide which paragraph mentions the following information? Write the number of the paragraph before each sentence.

_____ 1. Despite extensive research in modern agricultural methods, there remains a shortfall in studies specifically focused on their cybersecurity.

_____ 2. Investigations have shown that certain agricultural technologies can interfere with the functionality of IoT systems.

_____ 3. The necessity for enhanced agricultural productivity is critical due to the expected rise in global population and hunger.

_____ 4. Various components of smart farms, from field operations to data management, face distinct types of security vulnerabilities.

_____ 5. Modern farming techniques combine growth with smarter, more efficient operations using advanced technology.

_____ 6. The advancement of digital technologies in agriculture introduces increased risks of cybersecurity threats.

_____ 7. The protection of physically distributed and lightly supervised IoT devices in agriculture poses a significant challenge.

_____ 8. Specific security issues arise from the unique nature of IoT implementations in agriculture, which need dedicated strategies to address.

_____ 9. While promising security solutions exist, they must be adapted and expanded to effectively protect the evolving landscape of smart farms.

_____ 10. Challenges in smart agriculture security are often heightened by inadequate network infrastructure, particularly in less urbanized regions.

D. Cloze

The information below is a summary of the text. Complete the summary by filling in the blanks with the words provided.

A. datasets	F. future
B. agriculture	G. disrupt
C. signal	H. robust
D. security	I. leveraging
E. risks	J. physical

The article "Cybersecurity Report: 'Smart Farms' Are Hackable Farms" highlights the evolving landscape of (1) _____, which has increasingly integrated digital technologies such as the Internet, IoT, big data, cloud computing, and AI to boost productivity and sustainability. However, this shift toward 'smart agriculture' also introduces significant cybersecurity (2) _____. Despite the critical role of these technologies in addressing the (3) _____ food needs of a growing global population, the (4)_____ aspects specific to smart agricultural systems have been relatively under explored.

Xing Yang and his team from Nanjing Agricultural University point out that while the potential threats vary—from sensor failures in livestock monitoring to intrusions in greenhouse control systems—the consequences can significantly (5) _____ farming operations. The research emphasizes that the IoT devices used in agriculture often operate in remote and less supervised environments, making them vulnerable to (6) _____ and cyber threats. Additionally, problems like (7) _____ interference in rural areas can lead to errors such as false positioning of Unmanned Aerial Vehicles (UAVs).

To mitigate these risks, the researchers advocate for the adoption of (8) _____ security measures including improved authentication protocols, privacy frameworks, and advanced intrusion detection systems. They also see potential in (9) _____ AI and blockchain to address some vulnerabilities, though they recognize that these technologies themselves can introduce new security challenges. Moreover, the lack of (10) _____ specific to smart agriculture for training AI models indicates a gap that must be filled to enhance the security of these systems effectively.

Language Building

A. Glossary

Proper Nouns
IoT（Internet of Things） 物联网，是指通过互联网连接和通信的物理设备和对象的网络。 **Nanjing Agricultural University** 南京农业大学 **IEEE/CAA Journal of Automatica Sinica** 自动化学报

Unmanned Aerial Vehicle（UAV）
无人飞行器，又称无人机，是利用无线电遥控设备和自备的程序控制装置操纵的，或者由机载计算机完全或间歇自主操纵的不载人飞机。

Sybil attacks
女巫攻击。Douceur 首次给出了女巫攻击的概念，即在对等网络中，单一节点具有多个身份标识，通过控制系统的大部分节点来削弱冗余备份的作用。

Zigbee
ZigBee 是一种低速短距离传输的无线上网协议，底层是采用 IEEE 802.15.4 标准的媒体访问层与物理层。其主要特色有低速、低耗电、低成本、低复杂度、快速、可靠、安全、支持大量网上节点、支持多种网上拓扑。

AI algorithm
人工智能算法

Academic Words	
cybersecurity (n.)	网络安全，网际安全
hackable (adj.)	可破解的
vulnerable (adj.)	脆弱的；易受伤害的
dub (v.)	戏称为；称为
artificial (a.)	人造的；人工的；人为的
project (v.)	预测；预计
acute (adj.)	严重的；(疾病)急性的；灵敏的
dramatically (adv.)	戏剧性地；戏剧地；显著地
prospect (n.)	展望；前景
malnutrition (n.)	营养不良
extensively (adv.)	广大地；广泛地
conventional (adj.)	传统的；习惯的
characteristic (n.)	特征；特点
enumerate (v.)	列举；枚举
countermeasure (n.)	对策；对抗手段；反措施
malfunction (n.)	故障；失灵；功能障碍
poultry (n.)	家禽；家禽的肉
livestock (n.)	牲畜；家畜
acquisition (n.)	获得，得到

Chapter 8　Smart Farm

malicious (adj.)	恶意的；恶毒的
insecticidal (adj.)	杀虫的
authentication (n.)	身份验证；认证
protocol (n.)	协议；议定书
robust (adj.)	强健的；强劲的；强壮的
cryptography (n.)	密码学；密码术
dataset (n.)	资料组
loophole (n.)	漏洞
node (n.)	节点

B. Words and Phrases

Exercises 1　Word Choice

Use the words in the box to finish the sentences.

vulnerability	malicious	enumerate	countermeasure	acquisition
robust	protocol	extensively	acute	artificial

1. The names of items are too numerous to _____.
2. Knowledge _____ is to students as production is to workers.
3. A British Foreign Office minister has made a _____ defense of the agreement.
4. Depression is the result of a genetic _____.
5. Some _____ rumors are circulating about his past

Exercise 2　Phrases

Match the words provided below with appropriate one in the box.

1. _____ attacks
2. Intrusion _____ system
3. _____ countermeasures
4. greenhouse _____
5. data _____ sensor
6. _____ vulnerability

cybersecurity
enumerate
malicious
acquisition
cultivation
detection

Exercise 3　Sentence Completion

Complete the sentences by filling in the blanks with phrases in the above exercise.

1. China _____ against restrictions on Chinese media in the Chaptered States.

2. _____ can do a lot of damage to websites.

3. Regulators were concerned that a Chinese company would gain access to sensitive _____ information held by 3Com's security division.

4. _____ can increase crop yield effectively.

5. Practice shows that the node can realize the wireless transmission of _____, and the definition and application of TEDS data.

6. _____ can do a good job of undercutting malicious attacks.

Exercise 4 Translation

Translate the sentences by using the words and phrases you have learned in the above two exercises.

1. 世界贸易组织在 2019 年允许华盛顿对欧洲价值 75 亿美元的出口采取反制措施。
_____

2. 大量恶意攻击使得网站停止运营。
_____

3. 温室栽培产业的发展已经成为衡量现代农业发展水平的重要标志之一。
_____

4. 通过对以往入侵检测系统的分析和比较，我们提出了一种改进后的分布式结构。
_____

5. 本文主要研究以 CCD 为主要数据采集传感器，以机器视觉识别算法为主要算法，以集成 DSP 功能的嵌入式 CPU 为硬件处理平台的汽车前方路况视觉识别系统。
_____

C. Collocation

Exercise 1 Modifiers

Find out the adjectives that modify the verb "increase" in the article. The first letter has been provided.

increase y____ (para. 1)

increase e____ (para. 3)

increase r____ (para. 8)

increase s____ (para. 1)

increase a____ (para. 6)

Chapter 8   Smart Farm

Exercise 2   Blank Filling

Scan the text and complete the sentences containing the word "increase". Complete the following sentences with the Chinese translations.

1. The nature of the packing material also can increase＿＿＿＿＿（传播疾病的风险）.

2. Excessive fertilization can not increase the＿＿＿＿（产量）significantly, but has negative effect on profit.

3. He predicted that the company could increase＿＿＿＿（利润率）by 0.8% this year and next.

4. The pesticide application system could reduce the impact of pesticides on environments, increase the＿＿＿＿（生态系统可持续性），and protect the operators.

5. The agricultural expert system is to increase＿＿＿＿（农业生产效率）and to promote

6. agricultural knowledge.

7. If farmers switch to grazing practices that mimic the movements of wild herds, this can easily increase＿＿＿＿（碳的比例）stored in dry soils from 1% to 2%.

8. To raise grain yield, it is necessary to increase＿＿＿＿（农业投入）and to ensure proper scale of growing area under grain crops.

9. Personal computers and the Internet promised to create wealth, increase＿＿＿（获取信息途径），and foster communication among their users.

Exercise 3   Translation

Translate the sentences below from Chinese to English using "increase" and its collocation in this section.

1. 饥饿和粮食不安全问题具有全球性，在一些地区甚至<u>急剧增加</u>。

_____

2. 体育参与率在1995年至2003年保持稳定，之后开始增加，但仅<u>略有增加</u>。

_____

3. 受贿罪是发生率最高的罪，并有<u>逐年递增</u>的趋势。

_____

4. 必须采取系统科学的措施，<u>稳步增加</u>农民收入。

_____

5. 火电是耗煤量最大的方式，近年来，能源日益紧张，煤炭价格也在<u>不断上涨</u>，因此，如何提高热效率以达到节煤的目的，成为众多电厂最关心的问题。

_____

D. Terminology

Exercise 1 Table Filling

Read the article and find the English technical terms according to the Chinese equivalents.

English Technical Terms	Chinese Equivalents
	网络安全漏洞
	列举对策
	恶意攻击
	数据采集传感器
	温室栽培
	入侵检测系统

Exercises 2 Blank Filling

Use the terms in the above exercise to complete the sentences below.

1. Yet with so much digital technology, naturally, also comes heightened potential—_____.

2. Agricultural IoT applications have unique characteristics that give rise to security issues, which the authors _____.

3. Plus, there are threats to data acquisition technologies, such as _____.

4. They found that the lamp's high voltage pulse affects the data transmission from Zigbee-based IoT devices and _____.

5. _____ could face control system intrusions.

6. The study also summarizes existing security and privacy countermeasures suitable for smart agriculture, including robust _____ and so on.

Critical Reading and Writing

A. Brainstorming

Work in groups. Fill in the table according to the instruction.

What are the potential implications, both positive and negative, of widespread adoption of smart agricultural practices on global food security and environmental sustainability? List your answers in the table below.

Positive Impacts	Negative Impacts
1	1
2	2
3	3

B. Critical Reading

Exercise 1: The following are the answers to the question provided by ChatGPT. Decide which of the following statements are positive impacts and which ones are negative ones.

Positive impacts: _____

Negative impacts: _____

> a. Dependency on technology, which could exacerbate disparities between large-scale industrial farms and smaller, resource-constrained farms.
>
> b. Increased food production efficiency, leading to higher yields and reduced resource waste.
>
> c. Enhanced ability to monitor and manage crop health, resulting in better pest and disease control.
>
> d. Concerns about data privacy and ownership, particularly regarding sensitive agricultural data collected by smart farm systems.
>
> e. Opportunities for precision farming, allowing farmers to optimize inputs such as water, fertilizer, and pesticides.
>
> f. Potential job displacement in rural communities as automation and AI technologies replace manual labor tasks traditionally performed by farm workers.
>
> g. Potential for sustainable agriculture practices, such as regenerative farming and organic farming, to be more effectively implemented.
>
> h. Environmental risks associated with increased reliance on digital infrastructure, such as e-waste generation and energy consumption

Exercise 2: Which of the ideas provided by ChatGPT you haven't thought of? Do you think ChatGPT helps you to expand your ideas? Why or why not?

C. Essay Writing

Having engaged in discussions and vocabulary preparation, you've likely generated numerous insightful ideas. Now, it's time to reflect on these ideas and the knowledge you've acquired by crafting an essay. Use the following instructions to guide your writing process:

 Topic

The Impact of Smart Farms on ... (a specific industry)

Background Information

In recent years, agriculture has undergone a transformative evolution with the integration of digital technologies into traditional farming practices. This phenomenon, often referred to as "smart farming" or "precision agriculture," leverages advancements in areas such as the Internet of Things (IoT), big data analytics, Artificial Intelligence (AI), and automation to revolutionize the way crops are grown, monitored, and managed. Smart farms hold the promise of increased productivity, sustainability, and efficiency in food production, addressing the challenges posed by a growing global population and environmental pressures.

Instructions

In an essay of approximately 300-450 words, explain why smart farms are becoming increasingly important in modern agriculture. Provide specific evidence and arguments to support your opinion.

Your essay should include the following components:

Introduction (approximately 50-75 words): Briefly introduce the topic and provide context for your analysis.

- State your thesis or main argument regarding the benefits of smart farms.

Body Paragraphs (approximately 250-350 words): Present your argument in detail, supported by specific evidence and examples in 3 separate paragraphs.

- Analyze the benefit in three aspects: increased efficiency, productivity, and sustainability.
- Provide case studies or examples of successful smart farm implementations.
- Analyze the role of data analytics, IoT, AI, and other emerging technologies in smart farms.

Conclusion (approximately 50-75 words): Summarize your main points and restate your thesis in light of the evidence presented.

- Summarize the key points and arguments presented in the essay.
- Reinforce the importance of smart farms in shaping the future of agriculture.

* Ensure that your essay is well-structured, logically organized, and supported by evidence from reputable sources. Use clear and concise language, and proofread your work carefully for grammar, punctuation, and spelling errors.

Discussion and Presentation

A. Group Discussion
Exercise 1: Think and discuss the questions below.

1. What are the main challenges faced by China in ensuring food security, and how do these challenges impact the global agricultural landscape?

2. How are digital and technological advancements transforming modern agriculture, particularly in China, and what are the potential benefits and drawbacks of these innovations?

3. In what ways can agricultural practices be made more sustainable and environmentally friendly, and why is this important for both local communities and global efforts to combat climate change?

Exercise 2: Read the article below. Search for any information related to the above questions. Do you think the answers provided in the article match yours? Share your ideas with the class.

B. Extended Reading-Smart Farm technology in China

Smart Farm Technology Can Transform Chinese Agriculture and Help Feed the Planet

Farmers play a vitally important role in the global community; their hard work and dedication produce the food and other essential crops that we all depend upon.

Agriculture is the most basic priority of a successful society. As the famous agronomist, Nobel Peace Prize winner and "father of the green revolution" Norman Borlaug said, "Everything else can wait; agriculture can't."

Among all the nations in the world, China has the largest population and thus the most mouths to feed. The well-being of our own people depends on our ability to feed ourselves with safe, nutritious food that is grown sustainably. This is not easy, given that we have 20 percent of the world population but only 7 percent of its arable land.

China's successful development over recent years has also contributed to a shortage of good farmland as well as a strain on water supplies. Beyond our borders, the whole world faces an urgent need to collaborate on climate change and environmental protection.

In 2020, the COVID-19 pandemic also disrupted farming, reminding us of the fragility of food supply chains and their critical importance to people everywhere. Holistically, it's plain to see that a strategic approach to farming is a key route to human well-being and stability.

China is meeting these challenges head on. Today, food security is the government's primary agricultural agenda, while key demands of Chinese consumers center around food safety and quality.

At the same time, Chinese policy makers fully understand that the future of China's agriculture sector lies in agricultural modernization, and the key to advancing agricultural modernization lies in the development of technology. All are crucial imperatives for China and therefore cited in the 14th Five-Year Plan.

Current agricultural policies are aimed at improving the quality and nutrition of crops, which means teaching farmers how to maintain yields without overusing fertilizer and pesticides. China has articulated a vision for the future of agriculture. We must improve the quality of arable lands through using high quality farming infrastructure and techniques.

This is why building a modern agricultural eco-system is important and where digital and technological platform can make a huge difference, one example being that we can now provide information on local weather patterns and weed and insect threats to enable farmers to confidently increase yields and reduce pesticide use.

With modern agriculture platform and smart farm technology including drone and satellite imagery and pattern modeling, we are able to provide farmers across the country the support they need, turning farmers' mobile phones into intelligent environmental tools and resources. These transformations are already taking place and will help Chinese farmers quickly leapfrog into modern, highly efficient agricultural techniques. At the same time, reduce the use of fertilizers, pesticides and save water.

With the help of agri-tech and new business model that is driven by digital innovation through entire agricultural value chain, small farmers can benefit from a huge and supportive ecosystem. And for the consumers they ultimately serve, the system can also solve their pain points.

Modern agriculture digital services enable full product traceability, giving consumers the ability to scan a code to see information on farm location, harvest date and sustainability. This is a premium service today but could be mainstream tomorrow.

Making agricultural production more sustainable and environmentally friendly, while digitally connecting farmers to the consumers who consume their food, not only represents the trend for Chinese agriculture, but also the future of global agricultural development.

We only have one planet. We need to pay more attention to the protection of soil health and water resources, to help life on our planet co-exist with growth and to protect biodiversity.

It is important to carry out and expand farmer training and capacity building projects all over the world, using scientific and technological research and development to help farmers quickly respond to the problems encountered in agricultural production, including how to combat climate change, abnormal weather, and other challenges.

Agriculture plays an important role in tackling global climate change issues. Facing the ambitious goal of achieving carbon neutrality, the agricultural sector and leading companies shoulder great responsibilities and face great opportunities.

Enabling farmers worldwide to be more productive, more environmentally sustainable, more independent, and more risk-resilient in the face of extreme weather and disease is the key to feeding and protecting the whole planet.

C. Presentation

After you have read the article, please choose one of the following topics to develop your ideas. Make a presentation with PowerPoint to the class.

Topic1 The Future of Agriculture in China: Navigating Challenges and Embracing Technological Innovation
- Explore the challenges faced by China in ensuring food security, including population growth, limited arable land, and environmental degradation.
- Discuss the role of technological innovations such as smart farms, digital platforms, and precision agriculture in addressing these challenges and advancing agricultural modernization.
- Highlight case studies and examples of successful implementation of digital and technological solutions in Chinese agriculture.

Topic 2 Sustainable Agriculture and Climate Change: Strategies for Global Food Security

- Focus on the importance of sustainable agricultural practices in addressing global challenges such as climate change and food security.
- Discuss the environmental impacts of conventional agricultural practices and the need for more sustainable approaches to farming.
- Explore strategies for promoting sustainable agriculture, including organic farming, agroforestry, and conservation agriculture.
- Examine the role of digital and technological innovations in supporting sustainable agriculture, such as precision farming and remote sensing technologies.

 Video

Video

A. Before You Watch

Read out the words below. Choose a word in the box to form an appropriate expression.

| agriculture   data   intelligence   spectrum   imagery   energy   work |

- capture_____
- data driven _____
- TV_____
- _____ together
- artificial_____
- solar_____producer
- satellite_____

B. While You Watch

Exercise 1: Discussion

Watch the video and discuss: How can the integration of technology, such as data analytics, artificial intelligence, and TV white spaces, contribute to addressing climate *change challenges in agriculture and promoting sustainability in the agricultural sector?*

Exercise 2: Dictation
Fill in the blanks with words and expressions you have heard from the video.

Agriculture is both a contributor to climate change. It will be affected by climate change and it could also be a solution to climate change. One of the most promising ways to address that problem is that of _____. That is, if you could_____ from the farm from different parts of the farm and then use _____ on top of that data to add value to be able to predict things that you otherwise cannot sense cannot measure, you can then use that to improve efficiencies in agriculture.

TV white spaces are unused_____ and they're really great when it comes to the agriculture scenario because they can travel very long distances.

A farmer needs to know what the farm looks like, not just what's about the soil also what's below the soil one way to create that view is if you can bring data from sensors.

If you can bring data from _____from drone imagery using artificial intelligence and _____, you can _____ to create views of the farm that you otherwise just could get with one drone flight.

Some of the TV space solutions could help solve their problem, the supply chain industry and the wind and_____ would love to know you know what the solar predictions or the wind predictions would be where some of the micro climate predictions to sort of help them solve that problem.

These are all fundamental computer science challenges where we as computer scientists need to _____ with the other scientists in the space so technology can help with climate adaptation and technology can also help with agriculture becoming _____.

📖 Summary and Reflection

Now that you have completed the chapter of smart farm, it's time to reflect on your learning and ensure you have met the goals set for the chapter. Follow these steps to complete the checklist:

- Carefully read through the checklist provided, which outlines the key learning objectives and goals of the smart farm chapter. For each item on the checklist, evaluate your own understanding and progress by checking the corresponding box.
- If you feel confident in your understanding and achievement of the goal, check the box; If you believe there are areas where you need further improvement or clarification, leave the box unchecked.

1. **Understanding of Smart Farm Technology:**

   ☐ Have I deepened my understanding of smart farm technology, including the underlying IoT devices, data analytics techniques, and precision agriculture methods?

   ☐ Can I explain the process of implementing smart farm technology clearly to others, including concepts such as sensor networks, data-driven decision-making, and automation?

2. **Critical Thinking Skills:**

   ☐ How effectively did I apply critical thinking skills to tasks such as analyzing the feasibility and implications of smart farm technology in different agricultural scenarios?

   ☐ Did I identify potential limitations or ethical considerations associated with the use of smart farm technology, such as data privacy concerns or environmental impacts?

3. **Reading Comprehension, Vocabulary, and Language Proficiency:**

   ☐ Have I improved my reading comprehension, vocabulary, and language proficiency through exercises focused on technical texts and terminology related to smart farm technology?

   ☐ Can I accurately define and use technical terms associated with smart farm technology in context?

4. **Collocating Words and Phrases:**

   ☐ Can I identify and analyze the collocation of words and phrases within the context of smart farm technology to enhance my understanding of technical terminology?

   ☐ Have I practiced using these collocations in sentences to reinforce my understanding?

5. **Understanding of Impacts:**
   □ How did I explore the potential impacts of smart farm technology on agricultural productivity, resource efficiency, and environmental sustainability?
   □ Can I articulate the broader societal implications of widespread adoption of smart farm practices, including economic, social, and environmental aspects?
6. **Presentation Skills:**
   □ How did I develop my presentation skills by creating and delivering presentations on topics related to smart farm technology, using tools like PowerPoint?
   □ Did I effectively organize information, use visual aids, and engage the audience to convey complex concepts related to smart farms?
7. **Reflection and Critical Analysis:**
   □ How did I reflect on and critically analyze the knowledge acquired throughout the chapter, particularly in relation to the ethical considerations and future applications of smart farms technology?
   □ Did I consider different perspectives and evaluate the significance of smart farm advancements in various agricultural sectors and global food security?